VOLUME NINETY NINE

ADVANCES IN
MARINE BIOLOGY
Special Volume on *Kogia* Biology: Part 2

ADVANCES IN MARINE BIOLOGY

Editorial Board

PROFESSOR BING CHEN
Memorial University of Newfoundland
bchen@mun.ca

PROFESSOR BERNHARD RIEGL
Nova University, Florida
rieglb@nova.edu

DR PAT HUTCHINGS
National Museum Australia
Pat.hutchings@austmus.gov.au

PROF FERDINANDO BOERO
Ecology, Universita Napoli
ferdinando.boero@unina.it

DR GOETZ REINICKE
German Oceanographic Museum, Deutsches Meeresmuseum
goetz.reinicke@meeresmuseum.de

VOLUME NINETY NINE

ADVANCES IN MARINE BIOLOGY

Special Volume on *Kogia* Biology: Part 2

Edited by

STEPHANIE PLÖN
BioConsult SH Husum, Germany

ACADEMIC PRESS
An imprint of Elsevier

Academic Press is an imprint of Elsevier
50 Hampshire Street, 5th Floor, Cambridge, MA 02139, United States
525 B Street, Suite 1650, San Diego, CA 92101, United States
125 London Wall, London, EC2Y 5AS, United Kingdom

First edition 2024

Copyright © 2024 Elsevier Ltd. All rights are reserved, including those for text and data mining, AI training, and similar technologies.

Publisher's note: Elsevier takes a neutral position with respect to territorial disputes or jurisdictional claims in its published content, including in maps and institutional affiliations.

No part of this publication may be reproduced or transmitted in any form or by any means, electronic or mechanical, including photocopying, recording, or any information storage and retrieval system, without permission in writing from the publisher. Details on how to seek permission, further information about the Publisher's permissions policies and our arrangements with organizations such as the Copyright Clearance Center and the Copyright Licensing Agency, can be found at our website: www.elsevier.com/permissions.

This book and the individual contributions contained in it are protected under copyright by the Publisher (other than as may be noted herein).

Notices
Knowledge and best practice in this field are constantly changing. As new research and experience broaden our understanding, changes in research methods, professional practices, or medical treatment may become necessary.

Practitioners and researchers must always rely on their own experience and knowledge in evaluating and using any information, methods, compounds, or experiments described herein. In using such information or methods they should be mindful of their own safety and the safety of others, including parties for whom they have a professional responsibility.

To the fullest extent of the law, neither the Publisher nor the authors, contributors, or editors, assume any liability for any injury and/or damage to persons or property as a matter of products liability, negligence or otherwise, or from any use or operation of any methods, products, instructions, or ideas contained in the material herein.

ISBN: 978-0-443-29720-5
ISSN: 0065-2881

For information on all Academic Press publications
visit our website at https://www.elsevier.com/books-and-journals

Publisher: Zoe Kruze
Acquisitions Editor: Jason Mitchell
Editorial Project Manager: Palash Sharma
Production Project Manager: Maria Shalini
Cover Designer: Gopalakrishnan Venkatraman

Typeset by MPS Limited, India

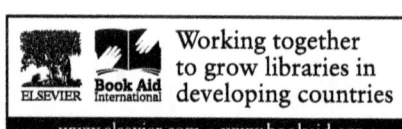

Contents

Contributors to Volume 99   vii
Series Contents for Last Fifteen Years   ix

1. Habitat suitability, occurrence, and behavior of dwarf sperm whales (*Kogia sima*) off St. Vincent and the Grenadines, Eastern Caribbean   1
Jeremy J. Kiszka, Guilherme Maricato, and Michelle Caputo

   1. Introduction   2
   2. Materials and methods   4
     2.1 Data collection   4
     2.2 Data analysis   5
   3. Results   8
     3.1 Occurrence and behavior   8
     3.2 Habitat suitability   8
     3.3 Photo-identification   9
   4. Discussion   11
   Acknowledgments   15
   Appendix A. Supporting information   16
   References   16

2. Strandings and at sea observations reveal the canary archipelago as an important habitat for pygmy and dwarf sperm whale   21
Vidal Martín, Marisa Tejedor, Manuel Carrillo, Mónica Pérez-Gil, Manuel Arbelo, Antonella Servidio, Enrique Pérez-Gil, Nuria Varo-Cruz, Francesca Fusar Poli, Sol Aliart, Gustavo Tejera, Marta Lorente, and Antonio Fernández

   1. Introduction   22
   2. Material and methods   25
     2.1 Study area   25
     2.2 Strandings   25
     2.3 Sightings at sea   26
     2.4 Species identification   27
   3. Results   28
     3.1 Strandings   28
     3.2 Sightings   38

4. Discussion 47
   4.1 Strandings 47
   4.2 Reproduction 49
   4.3 Presence in the Canary Archipelago 50
   4.4 Predation 52
   4.5 Conservation 54
5. Conclusions 56
6. Future perspectives 57
Acknowledgments 58
References 58

3. Records from visual surveys, strandings and eDNA sampling reveal the regular use of Reunion waters by dwarf sperm whales 65

Violaine Dulau, Vanessa Estrade, Aymeric Bein, Natacha Nikolic, Adrian Fajeau, Jean-Marc Gancille, Julie Martin, Emmanuelle Leroy, and Jean-Sebastien Philippe

1. Introduction 66
2. Material and methods 69
   2.1 Boat-based and aerial surveys 69
   2.2 Environmental DNA (eDNA) 71
   2.3 Strandings 73
3. Results 74
   3.1 Distribution surveys 74
   3.2 Environmental DNA 77
   3.3 Strandings 78
4. Discussion 85
   4.1 Species occurrence 86
   4.2 Spatial distribution around Reunion 87
   4.3 Habitat use 89
   4.4 Group size and biology 90
   4.5 Threats 92
5. Conclusions 93
Acknowledgments 94
Appendix A. Supporting information 94
References 94

Contributors to Volume 99

Sol Aliart
Society for the Study of Cetacean in the Canary Archipelago (SECAC), Canary Islands Cetacean Research Centre, Canary Islands Stranding Network, Lanzarote, Canary Islands, Spain

Manuel Arbelo
Veterinary Histology and Pathology, Atlantic Center for Cetacean Research, University Institute of Animal Health and Food Safety (IUSA), Veterinary School, University of Las Palmas de Gran Canaria, Canary Islands, Spain

Aymeric Bein
BIOTOPE, Saint André, Reunion

Michelle Caputo
Institute of Environment, Department of Biological Sciences, Florida International University, North Miami, United States

Manuel Carrillo
Canarias Conservación, La Laguna, Tenerife, Spain

Violaine Dulau
GLOBICE-Reunion, Grand-Bois, Saint Pierre, Reunion

Vanessa Estrade
GLOBICE-Reunion, Grand-Bois, Saint Pierre, Reunion

Adrian Fajeau
GLOBICE-Reunion, Grand-Bois, Saint Pierre, Reunion

Antonio Fernández
Veterinary Histology and Pathology, Atlantic Center for Cetacean Research, University Institute of Animal Health and Food Safety (IUSA), Veterinary School, University of Las Palmas de Gran Canaria, Canary Islands, Spain

Francesca Fusar Poli
Society for the Study of Cetacean in the Canary Archipelago (SECAC), Canary Islands Cetacean Research Centre, Canary Islands Stranding Network, Lanzarote, Canary Islands, Spain

Jean-Marc Gancille
GLOBICE-Reunion, Grand-Bois, Saint Pierre, Reunion

Jeremy J. Kiszka
Institute of Environment, Department of Biological Sciences, Florida International University, North Miami, United States

Emmanuelle Leroy
GLOBICE-Reunion, Grand-Bois, Saint Pierre, Reunion

Marta Lorente
Society for the Study of Cetacean in the Canary Archipelago (SECAC), Canary Islands Cetacean Research Centre, Canary Islands Stranding Network, Lanzarote, Canary Islands, Spain

Guilherme Maricato
Institute of Environment, Department of Biological Sciences, Florida International University, North Miami, United States

Julie Martin
GLOBICE-Reunion, Grand-Bois, Saint Pierre, Reunion

Vidal Martín
Society for the Study of Cetacean in the Canary Archipelago (SECAC), Canary Islands Cetacean Research Centre, Canary Islands Stranding Network, Lanzarote, Canary Islands, Spain

Natacha Nikolic
Agence de Recherche pour la Biodiversité à la Réunion (ARBRE), Saint Gilles, Reunion; INRAE, ECOBIOP, AQUA, Saint-Pee-sur-Nivelle, France

Jean-Sebastien Philippe
BIOTOPE, Saint André, Reunion

Enrique Pérez-Gil
Cetacean and Marine Research Institute of the Canary Islands (CEAMAR), Playa Honda, Lanzarote, Canary Islands, Spain

Mónica Pérez-Gil
Cetacean and Marine Research Institute of the Canary Islands (CEAMAR), Playa Honda, Lanzarote, Canary Islands, Spain

Antonella Servidio
Cetacean and Marine Research Institute of the Canary Islands (CEAMAR), Playa Honda, Lanzarote, Canary Islands, Spain

Marisa Tejedor
Canary Islands Cetaceans Stranding Network, Playa Blanca, Lanzarote, Canary Islands, Spain

Gustavo Tejera
Canary Islands' Ornithology and Natural History Group (GOHNIC)

Nuria Varo-Cruz
Cetacean and Marine Research Institute of the Canary Islands (CEAMAR), Playa Honda, Lanzarote, Canary Islands, Spain

Series Contents for Last Fifteen Years[*,†]

Volume 44, 2003.
Hirst, A. G., Roff, J. C. and Lampitt, R. S. A synthesis of growth rates in epipelagic invertebrate zooplankton. pp. 3–142.
Boletzky, S. von. Biology of early life stages in cephalopod molluscs. pp. 143–203.
Pittman, S. J. and McAlpine, C. A. Movements of marine fish and decapod crustaceans: process, theory and application. pp. 205–294.
Cutts, C. J. Culture of harpacticoid copepods: potential as live feed for rearing marine fish. pp. 295–315.

Volume 45, 2003.
Cumulative Taxonomic and Subject Index.

Volume 46, 2003.
Gooday, A. J. Benthic foraminifera (Protista) as tools in deep-water palaeoceanography: environmental influences on faunal characteristics. pp. 1–90.
Subramoniam, T. and Gunamalai, V. Breeding biology of the intertidal sand crab, Emerita (Decapoda: Anomura). pp. 91–182.
Coles, S. L. and Brown, B. E. Coral bleaching—capacity for acclimatization and adaptation. pp. 183–223.
Dalsgaard J., St. John M., Kattner G., Müller-Navarra D. and Hagen W. Fatty acid trophic markers in the pelagic marine environment. pp. 225–340.

Volume 47, 2004.
Southward, A. J., Langmead, O., Hardman-Mountford, N. J., Aiken, J., Boalch, G. T., Dando, P. R., Genner, M. J., Joint, I., Kendall, M. A., Halliday, N. C., Harris, R. P., Leaper, R., Mieszkowska, N., Pingree, R. D., Richardson, A. J., Sims, D.W., Smith, T., Walne, A. W. and Hawkins, S. J. Long-term oceanographic and ecological research in the western English Channel. pp. 1–105.
Queiroga, H. and Blanton, J. Interactions between behaviour and physical forcing in the control of horizontal transport of decapod crustacean larvae. pp. 107–214.

[*] The full list of contents for volumes 1–37 can be found in volume 38
[†] The full list of contents for volumes 38–43 can be found in volume 78

Braithwaite, R. A. and McEvoy, L. A. Marine biofouling on fish farms and its remediation. pp. 215–252.

Frangoulis, C., Christou, E. D. and Hecq, J. H. Comparison of marine copepod outfluxes: nature, rate, fate and role in the carbon and nitrogen cycles. pp. 253–309.

Volume 48, 2005.
Canfield, D. E., Kristensen, E. and Thamdrup, B. Aquatic Geomicrobiology. pp. 1–599.

Volume 49, 2005.
Bell, J. D., Rothlisberg, P. C., Munro, J. L., Loneragan, N. R., Nash, W. J., Ward, R. D. and Andrew, N. L. Restocking and stock enhancement of marine invertebrate fisheries. pp. 1–358.

Volume 50, 2006.
Lewis, J. B. Biology and ecology of the hydrocoral *Millepora* on coral reefs. pp. 1–55.

Harborne, A. R., Mumby, P. J., Micheli, F., Perry, C. T., Dahlgren, C. P., Holmes, K. E., and Brumbaugh, D. R. The functional value of Caribbean coral reef, seagrass and mangrove habitats to ecosystem processes. pp. 57–189.

Collins, M. A. and Rodhouse, P. G. K. Southern ocean cephalopods. pp. 191–265.

Tarasov, V. G. Effects of shallow-water hydrothermal venting on biological communities of coastal marine ecosystems of the western Pacific. pp. 267-410.

Volume 51, 2006.
Elena Guijarro Garcia. The fishery for Iceland scallop (*Chlamys islandica*) in the Northeast Atlantic. pp. 1–55.

Jeffrey, M. Leis. Are larvae of demersal fishes plankton or nekton? pp. 57–141.

John C. Montgomery, Andrew Jeffs, Stephen D. Simpson, Mark Meekan and Chris Tindle. Sound as an orientation cue for the pelagic larvae of reef fishes and decapod crustaceans. pp. 143–196.

Carolin E. Arndt and Kerrie M. Swadling. Crustacea in Arctic and Antarctic sea ice: Distribution, diet and life history strategies. pp. 197–315.

Volume 52, 2007.
Leys, S. P., Mackie, G. O. and Reiswig, H. M. The Biology of Glass Sponges. pp. 1–145.
Garcia E. G. The Northern Shrimp (*Pandalus borealis*) Offshore Fishery in the Northeast Atlantic. pp. 147–266.
Fraser K. P. P. and Rogers A. D. Protein Metabolism in Marine Animals: The Underlying Mechanism of Growth. pp. 267–362.

Volume 53, 2008.
Dustin J. Marshall and Michael J. Keough. The Evolutionary Ecology of Offspring Size in Marine Invertebrates. pp. 1–60.
Kerry A. Naish, Joseph E. Taylor III, Phillip S. Levin, Thomas P. Quinn, James R. Winton, Daniel Huppert, and Ray Hilborn. An Evaluation of the Effects of Conservation and Fishery Enhancement Hatcheries on Wild Populations of Salmon. pp. 61–194.
Shannon Gowans, Bernd Würsig, and Leszek Karczmarski. The Social Structure and Strategies of Delphinids: Predictions Based on an Ecological Framework. pp. 195–294.

Volume 54, 2008.
Bridget S. Green. Maternal Effects in Fish Populations. pp. 1–105.
Victoria J. Wearmouth and David W. Sims. Sexual Segregation in Marine Fish, Reptiles, Birds and Mammals: Behaviour Patterns, Mechanisms and Conservation Implications. pp. 107–170.
David W. Sims. Sieving a Living: A Review of the Biology, Ecology and Conservation Status of the Plankton-Feeding Basking Shark *Cetorhinus maximus*. pp. 171–220.
Charles H. Peterson, Kenneth W. Able, Christin Frieswyk DeJong, Michael F. Piehler, Charles A. Simenstad, and Joy B. Zedler. Practical Proxies for Tidal Marsh Ecosystem Services: Application to Injury and Restoration. pp. 221–266.

Volume 55, 2008.
Annie Mercier and Jean-Francois Hamel. Introduction. pp. 1–6.
Annie Mercier and Jean-Francois Hamel. Gametogenesis. pp. 7–72.
Annie Mercier and Jean-Francois Hamel. Spawning. pp. 73–168.
Annie Mercier and Jean-Francois Hamel. Discussion. pp. 169–194.

Volume 56, 2009.

Philip C. Reid, Astrid C. Fischer, Emily Lewis-Brown, Michael P. Meredith, Mike Sparrow, Andreas J. Andersson, Avan Antia, Nicholas R. Bates, Ulrich Bathmann, Gregory Beaugrand, Holger Brix, Stephen Dye, Martin Edwards, Tore Furevik, Reidun Gangst, Hjalmar Hatun, Russell R. Hopcroft, Mike Kendall, Sabine Kasten, Ralph Keeling, Corinne Le Quere, Fred T. Mackenzie, Gill Malin, Cecilie Mauritzen, Jon Olafsson, Charlie Paull, Eric Rignot, Koji Shimada, Meike Vogt, Craig Wallace, Zhaomin Wang and Richard Washington. Impacts of the Oceans on Climate Change. pp. 1–150.

Elvira S. Poloczanska, Colin J. Limpus and Graeme C. Hays. Vulnerability of Marine Turtles to Climate Change. pp. 151–212.

Nova Mieszkowska, Martin J. Genner, Stephen J. Hawkins and David W. Sims. Effects of Climate Change and Commercial Fishing on Atlantic Cod *Gadus morhua*. pp. 213–274.

Iain C. Field, Mark G. Meekan, Rik C. Buckworth and Corey J. A. Bradshaw. Susceptibility of Sharks, Rays and Chimaeras to Global Extinction. pp. 275–364.

Milagros Penela-Arenaz, Juan Bellas and Elsa Vazquez. Effects of the *Prestige* Oil Spill on the Biota of NW Spain: 5 Years of Learning. pp. 365–396.

Volume 57, 2010.

Geraint A. Tarling, Natalie S. Ensor, Torsten Fregin, William P. Good-all-Copestake and Peter Fretwell. An Introduction to the Biology of Northern Krill (*Meganyctiphanes norvegica* Sars). pp. 1–40.

Tomaso Patarnello, Chiara Papetti and Lorenzo Zane. Genetics of Northern Krill (*Meganyctiphanes norvegica* Sars). pp. 41–58.

Geraint A. Tarling. Population Dynamics of Northern Krill (*Meganyctiphanes norvegica* Sars). pp. 59–90.

John I. Spicer and Reinhard Saborowski. Physiology and Metabolism of Northern Krill (*Meganyctiphanes norvegica* Sars). pp. 91–126.

Katrin Schmidt. Food and Feeding in Northern Krill (*Meganyctiphanes norvegica* Sars). pp. 127–172.

Friedrich Buchholz and Cornelia Buchholz. Growth and Moulting in Northern Krill (*Meganyctiphanes norvegica* Sars). pp. 173–198.

Janine Cuzin-Roudy. Reproduction in Northern Krill. pp. 199–230.

Edward Gaten, Konrad Wiese and Magnus L. Johnson. Laboratory-Based Observations of Behaviour in Northern Krill (*Meganyctiphanes norvegica* Sars). pp. 231–254.

Stein Kaartvedt. Diel Vertical Migration Behaviour of the Northern Krill (*Meganyctiphanes norvegica* Sars). pp. 255–276.

Yvan Simard and Michel Harvey. Predation on Northern Krill (*Meganyctiphanes norvegica* Sars). pp. 277–306.

Volume 58, 2010.

A. G. Glover, A. J. Gooday, D. M. Bailey, D. S. M. Billett, P. Chevaldonné, A. Colaço, J. Copley, D. Cuvelier, D. Desbruyères, V. Kalogeropoulou, M. Klages, N. Lampadariou, C. Lejeusne, N. C. Mestre, G. L. J. Paterson, T. Perez, H. Ruhl, J. Sarrazin, T. Soltwedel, E. H. Soto, S. Thatje, A. Tselepides, S. Van Gaever, and A. Vanreusel. Temporal Change in Deep-Sea Benthic Ecosystems: A Review of the Evidence From Recent Time-Series Studies. pp. 1–96.

Hilario Murua. The Biology and Fisheries of European Hake, *Merluccius merluccius*, in the North-East Atlantic. pp. 97–154.

Jacopo Aguzzi and Joan B. Company. Chronobiology of Deep-Water Decapod Crustaceans on Continental Margins. pp. 155–226.

Martin A. Collins, Paul Brickle, Judith Brown, and Mark Belchier. The Patagonian Toothfish: Biology, Ecology and Fishery. pp. 227–300.

Volume 59, 2011.

Charles W. Walker, Rebecca J. Van Beneden, Annette F. Muttray, S. Anne Böttger, Melissa L. Kelley, Abraham E. Tucker, and W. Kelley Thomas. p53 Superfamily Proteins in Marine Bivalve Cancer and Stress Biology. pp 1–36.

Martin Wahl, Veijo Jormalainen, Britas Klemens Eriksson, James A. Coyer, Markus Molis, Hendrik Schubert, Megan Dethier, Anneli Ehlers, Rolf Karez, Inken Kruse, Mark Lenz, Gareth Pearson, Sven Rohde, Sofia A. Wikström, and Jeanine L. Olsen. Stress Ecology in *Fucus*: Abiotic, Biotic and Genetic Interactions. pp. 37–106.

Steven R. Dudgeon and Janet E. Kübler. Hydrozoans and the Shape of Things to Come. pp. 107–144.

Miles Lamare, David Burritt, and Kathryn Lister. Ultraviolet Radiation and Echinoderms: Past, Present and Future Perspectives. pp. 145–187.

Volume 60, 2011.

Tatiana A. Rynearson and Brian Palenik. Learning to Read the Oceans: Genomics of Marine Phytoplankton. pp. 1–40.

Les Watling, Scott C. France, Eric Pante and Anne Simpson. Biology of Deep-Water Octocorals. pp. 41–122.

Cristián J. Monaco and Brian Helmuth. Tipping Points, Thresholds and the Keystone Role of Physiology in Marine Climate Change Research. pp. 123–160.

David A. Ritz, Alistair J. Hobday, John C. Montgomery and Ashley J.W. Ward. Social Aggregation in the Pelagic Zone with Special Reference to Fish and Invertebrates. pp. 161–228.

Volume 61, 2012.

Gert Wörheide, Martin Dohrmann, Dirk Erpenbeck, Claire Larroux, Manuel Maldonado, Oliver Voigt, Carole Borchiellini and Denis Lavrov. Deep Phylogeny and Evolution of Sponges (Phylum Porifera). pp. 1–78.

Paco Cárdenas, Thierry Pérez and Nicole Boury-Esnault. Sponge Systematics Facing New Challenges. pp. 79–210.

Klaus Rützler. The Role of Sponges in the Mesoamerican Barrier-Reef Ecosystem, Belize. pp. 211–272.

Janie Wulff. Ecological Interactions and the Distribution, Abundance, and Diversity of Sponges. pp. 273–344.

Maria J. Uriz and Xavier Turon. Sponge Ecology in the Molecular Era. pp. 345–410.

Volume 62, 2012.

Sally P. Leys and April Hill. The Physiology and Molecular Biology of Sponge Tissues. pp. 1–56.

Robert W. Thacker and Christopher J. Freeman. Sponge-Microbe Symbioses: Recent Advances and New Directions. pp. 57–112.

Manuel Maldonado, Marta Ribes and Fleur C. van Duyl. Nutrient Fluxes Through Sponges: Biology, Budgets, and Ecological Implications. pp. 113–182.

Gregory Genta-Jouve and Olivier P. Thomas. Sponge Chemical Diversity: From Biosynthetic Pathways to Ecological Roles. pp. 183–230.

Xiaohong Wang, Heinz C. Schröder, Matthias Wiens, Ute Schloßmacher and Werner E. G. Müller. Biosilica: Molecular Biology, Biochemistry and Function in Demosponges as well as its Applied Aspects for Tissue Engineering. pp. 231–272.

Klaske J. Schippers, Detmer Sipkema, Ronald Osinga, Hauke Smidt, Shirley A. Pomponi, Dirk E. Martens and René H. Wijffels. Cultivation of Sponges, Sponge Cells and Symbionts: Achievements and Future Prospects. pp. 273–338.

Volume 63, 2012.
Michael Stat, Andrew C. Baker, David G. Bourne, Adrienne M. S. Correa, Zac Forsman, Megan J. Huggett, Xavier Pochon, Derek Skillings, Robert J. Toonen, Madeleine J. H. van Oppen, and Ruth D. Gates. Molecular Delineation of Species in the Coral Holobiont. pp. 1–66.
Daniel Wagner, Daniel G. Luck, and Robert J. Toonen. The Biology and Ecology of Black Corals (Cnidaria: Anthozoa: Hexacorallia: Antipatharia). pp. 67–132.
Cathy H. Lucas, William M. Graham, and Chad Widmer. Jellyfish Life Histories: Role of Polyps in Forming and Maintaining Scyphomedusa Populations. pp. 133–196.
T. Aran Mooney, Maya Yamato, and Brian K. Branstetter. Hearing in Cetaceans: From Natural History to Experimental Biology. pp. 197–246.

Volume 64, 2013.
Dale Tshudy. Systematics and Position of *Nephrops* Among the Lobsters. pp. 1–26.
Mark P. Johnson, Colm Lordan, and Anne Marie Power. Habitat and Ecology of *Nephrops norvegicus*. pp. 27–64.
Emi Katoh, Valerio Sbragaglia, Jacopo Aguzzi, and Thomas Breithaupt. Sensory Biology and Behaviour of *Nephrops norvegicus*. pp. 65–106.
Edward Gaten, Steve Moss, and Magnus L. Johnson. The Reniform Reflecting Superposition Compound Eyes of *Nephrops norvegicus*: Optics, Susceptibility to Light-Induced Damage, Electrophysiology and a Ray Tracing Model. pp. 107–148.
Susanne P. Eriksson, Bodil Hernroth, and Susanne P. Baden. Stress Biology and Immunology in *Nephrops norvegicus*. pp. 149–200.
Adam Powell and Susanne P. Eriksson. Reproduction: Life Cycle, Larvae and Larviculture. pp. 201–246.
Anette Ungfors, Ewen Bell, Magnus L. Johnson, Daniel Cowing, Nicola C. Dobson, Ralf Bublitz, and Jane Sandell. *Nephrops* Fisheries in European Waters. pp. 247–314.

Volume 65, 2013.
Isobel S.M. Bloor, Martin J. Attrill, and Emma L. Jackson. A Review of the Factors Influencing Spawning, Early Life Stage Survival and Recruitment Variability in the Common Cuttlefish (*Sepia officinalis*). pp. 1–66.
Dianna K. Padilla and Monique M. Savedo. A Systematic Review of Phenotypic Plasticity in Marine Invertebrate and Plant Systems. pp. 67–120.

Leif K. Rasmuson. The Biology, Ecology and Fishery of the Dungeness crab, *Cancer magister*. pp. 121–174.

Volume 66, 2013.

Lisa-ann Gershwin, Anthony J. Richardson, Kenneth D. Winkel, Peter J. Fenner, John Lippmann, Russell Hore, Griselda Avila-Soria, David Brewer, Rudy J. Kloser, Andy Steven, and Scott Condie. Biology and Ecology of Irukandji Jellyfish (Cnidaria: Cubozoa). pp. 1–86.

April M. H. Blakeslee, Amy E. Fowler, and Carolyn L. Keogh. Marine Invasions and Parasite Escape: Updates and New Perspectives. pp. 87–170.

Michael P. Russell. Echinoderm Responses to Variation in Salinity. pp. 171–212.

Daniela M. Ceccarelli, A. David McKinnon, Serge Andréfouët, Valerie Allain, Jock Young, Daniel C. Gledhill, Adrian Flynn, Nicholas J. Bax, Robin Beaman, Philippe Borsa, Richard Brinkman, Rodrigo H. Bustamante, Robert Campbell, Mike Cappo, Sophie Cravatte, Stéphanie D'Agata, Catherine M. Dichmont, Piers K. Dunstan, Cécile Dupouy, Graham Edgar, Richard Farman, Miles Furnas, Claire Garrigue, Trevor Hutton, Michel Kulbicki, Yves Letourneur, Dhugal Lindsay, Christophe Menkes, David Mouillot, Valeriano Parravicini, Claude Payri, Bernard Pelletier, Bertrand Richer de Forges, Ken Ridgway, Martine Rodier, Sarah Samadi, David Schoeman, Tim Skewes, Steven Swearer, Laurent Vigliola, Laurent Wantiez, Alan Williams, Ashley Williams, and Anthony J. Richardson. The Coral Sea: Physical Environment, Ecosystem Status and Biodiversity Assets. pp. 213–290.

Volume 67, 2014.

Erica A.G. Vidal, Roger Villanueva, José P. Andrade, Ian G. Gleadall, José Iglesias, Noussithé Koueta, Carlos Rosas, Susumu Segawa, Bret Grasse, Rita M. Franco-Santos, Caroline B. Albertin, Claudia Caamal-Monsreal, Maria E. Chimal, Eric Edsinger-Gonzales, Pedro Gallardo, Charles Le Pabic, Cristina Pascual, Katina Roumbedakis, and James Wood. Cephalopod Culture: Current Status of Main Biological Models and Research Priorities. pp. 1–98.

Paul G.K. Rodhouse, Graham J. Pierce, Owen C. Nichols, Warwick H.H. Sauer, Alexander I. Arkhipkin, Vladimir V. Laptikhovsky, Marek R. Lipiński, Jorge E. Ramos, Michaël Gras, Hideaki Kidokoro, Kazuhiro Sadayasu, João Pereira, Evgenia Lefkaditou, Cristina Pita, Maria

Gasalla, Manuel Haimovici, Mitsuo Sakai, and Nicola Downey. Environmental Effects on Cephalopod Population Dynamics: Implications for Management of Fisheries. pp. 99–234.

Henk-Jan T. Hoving, José A.A. Perez, Kathrin Bolstad, Heather Braid, Aaron B. Evans, Dirk Fuchs, Heather Judkins, Jesse T. Kelly, Jośe E.A.R. Marian, Ryuta Nakajima, Uwe Piatkowski, Amanda Reid, Michael Vecchione, and José C.C. Xavier. The Study of Deep-Sea Cephalopods. pp. 235–362.

Jean-Paul Robin, Michael Roberts, Lou Zeidberg, Isobel Bloor, Almendra Rodriguez, Felipe Briceño, Nicola Downey, Maite Mascaró, Mike Navarro, Angel Guerra, Jennifer Hofmeister, Diogo D. Barcellos, Silvia A.P. Lourenço, Clyde F.E. Roper, Natalie A. Moltschaniwskyj, Corey P. Green, and Jennifer Mather. Transitions During Cephalopod Life History: The Role of Habitat, Environment, Functional Morphology and Behaviour. pp. 363–440.

Volume 68, 2014.

Paul K.S. Shin, Siu Gin Cheung, Tsui Yun Tsang, and Ho Yin Wai. Ecology of Artificial Reefs in the Subtropics. pp. 1–64.

Hrafnkell Eiríksson. Reproductive Biology of Female Norway Lobster, *Nephrops norvegicus* (Linnaeus, 1758) Leach, in Icelandic Waters During the Period 1960-2010: Comparative Overview of Distribution Areas in the Northeast Atlantic and the Mediterranean. pp. 65–210.

Volume 69, 2014.

Ray Hilborn. Introduction to Marine Managed Areas. pp. 1–14.

Philip N. Trathan, Martin A. Collins, Susie M. Grant, Mark Belchier, David K.A. Barnes, Judith Brown, and Iain J. Staniland. The South Georgia and the South Sandwich Islands MPA: Protecting A Biodiverse Oceanic Island Chain Situated in the Flow of the Antarctic Circumpolar Current. pp. 15-78.

Richard P. Dunne, Nicholas V.C. Polunin, Peter H. Sand, and Magnus L. Johnson. The Creation of the Chagos Marine Protected Area: A Fisheries Perspective. pp. 79–128.

Michelle T. Schärer-Umpierre, Daniel Mateos-Molina, Richard Appeldoorn, Ivonne Bejarano, Edwin A. Hernández-Delgado, Richard S. Nemeth, Michael I. Nemeth, Manuel Valdés-Pizzini, and Tyler B. Smith. Marine Managed Areas and Associated Fisheries in the US Caribbean. pp. 129–152.

Alan M. Friedlander, Kostantinos A. Stamoulis, John N. Kittinger, Jeffrey C. Drazen, and Brian N. Tissot. Understanding the Scale of

Marine Protection in Hawai'i: From Community-Based Management to the Remote Northwestern Hawaiian Islands. pp. 153–204.

Louis W. Botsford, J. Wilson White, Mark H. Carr, and Jennifer E. Caselle. Marine Protected Area Networks in California, USA. pp. 205–252.

Bob Kearney and Graham Farebrother. Inadequate Evaluation and Management of Threats in Australia's Marine Parks, Including the Great Barrier Reef, Misdirect Marine Conservation. pp. 253–288.

Randi Rotjan, Regen Jamieson, Ben Carr, Les Kaufman, Sangeeta Mangubhai, David Obura, Ray Pierce, Betarim Rimon, Bud Ris, Stuart Sandin, Peter Shelley, U. Rashid Sumaila, Sue Taei, Heather Tausig, Tukabu Teroroko, Simon Thorrold, Brooke Wikgren, Teuea Toatu, and Greg Stone. Establishment, Management, and Maintenance of the Phoenix Islands Protected Area. pp. 289–324.

Alex J. Caveen, Clare Fitzsimmons, Margherita Pieraccini, Euan Dunn, Christopher J. Sweeting, Magnus L. Johnson, Helen Bloomfield, Estelle V. Jones, Paula Lightfoot, Tim S. Gray, Selina M. Stead, and Nicholas V. C. Polunin. Diverging Strategies to Planning an Ecologically Coherent Network of MPAs in the North Sea: The Roles of Advocacy, Evidence and Pragmatism in the Face of Uncertaintya. pp. 325–370.

Carlo Pipitone, Fabio Badalamenti, Tomás Vega Fernández, and Giovanni D'Anna. Spatial Management of Fisheries in the Mediterranean Sea: Problematic Issues and a Few Success Stories. pp. 371–402.

Volume 70, 2015.
Alex D. Rogers, Christopher Yesson, and Pippa Gravestock. A Biophysical and Economic Profile of South Georgia and the South Sandwich Islands as Potential Large-Scale Antarctic Protected Areas. pp. 1–286.

Volume 71, 2015.
Ricardo Calado and Miguel Costa Leal. Trophic Ecology of Benthic Marine Invertebrates with Bi-Phasic Life Cycles: What Are We Still Missing? pp. 1–70.

Jesse M.A. van der Grient and Alex D. Rogers. Body Size Versus Depth: Regional and Taxonomical Variation in Deep-Sea Meio- and Macrofaunal Organisms. pp. 71–108.

Lorena Basso, Maite Vázquez-Luis, José R. García-March, Salud Deudero, Elvira Alvarez, Nardo Vicente, Carlos M. Duarte, and Iris E. Hendriks. The Pen Shell, *Pinna nobilis*: A Review of Population Status and Recommended Research Priorities in the Mediterranean Sea. pp. 109–160.

Volume 72, 2015.
Thomas A. Jefferson and Barbara E. Curry. Humpback Dolphins: A Brief Introduction to the Genus *Sousa*. pp. 1–16.
Sarah Piwetz, David Lundquist, and Bernd Würsig. Humpback Dolphin (Genus *Sousa*) Behavioural Responses to Human Activities. pp. 17–46.
Tim Collins. Re-assessment of the Conservation Status of the Atlantic Humpback Dolphin, *Sousa teuszii* (Kükenthal, 1892) Using the IUCN Red List Criteria. pp. 47–78.
Caroline R. Weir and Tim Collins. A Review of the Geographical Distribution and Habitat of the Atlantic Humpback Dolphin (*Sousa teuszii*). pp. 79–118.
Gill T Braulik, Ken Findlay, Salvatore Cerchio, and Robert Baldwin. Assessment of the Conservation Status of the Indian Ocean Humpback Dolphin (*Sousa plumbea*) Using the IUCN Red List Criteria. pp. 119–142.
Stephanie Plön, Victor G. Cockcroft, and William P. Froneman. The Natural History and Conservation of Indian Ocean Humpback Dolphins (*Sousa plumbea*) in South African Waters. pp. 143–162.
Salvatore Cerchio, Norbert Andrianarivelo, and Boris Andrianantenaina. Ecology and Conservation Status of Indian Ocean Humpback Dolphins (*Sousa plumbea*) in Madagascar. pp. 163–200.
Muhammad Shoaib Kiani and Koen Van Waerebeek. A Review of the Status of the Indian Ocean Humpback Dolphin *Sousa plumbea* in Pakistan. pp. 201–228.
Dipani Sutaria, Divya Panicker, Ketki Jog, Mihir Sule, Rahul Muralidharan, and Isha Bopardikar. Humpback Dolphins (Genus *Sousa*) in India: An Overview of Status and Conservation Issues. pp. 229–256.

Volume 73, 2016.
Thomas A. Jefferson and Brian D. Smith. Re-assessment of the Conservation Status of the Indo-Pacific Humpback Dolphin (*Sousa chinensis*) Using the IUCN Red List Criteria. pp. 1–26.
Leszek Karczmarski, Shiang-Lin Huang, Carmen K. M. Or, Duan Gui, Stephen C. Y. Chan, Wenzhi Lin, Lindsay Porter, Wai-Ho Wong, Ruiqiang Zheng, Yuen-Wa Ho, Scott Y. S. Chui, Angelico Jose C. Tiongson, Yaqian Mo, Wei-Lun Chang, John H. W. Kwok, Ricky W. K. Tang, Andy T. L. Lee, Sze-Wing Yiu, Mark Keith, Glenn Gailey, and Yuping Wu. Humpback Dolphins in Hong Kong and the Pearl River Delta: Status, Threats and Conservation Challenges. pp. 27–64.

Bernd Würsig, E.C.M. Parsons, Sarah Piwetz, and Lindsay Porter. The Behavioural Ecology of Indo-Pacific Humpback Dolphins in Hong Kong. pp. 65–90.

John Y. Wang, Kimberly N. Riehl, Michelle N. Klein, Shiva Javdan, Jordan M. Hoffman, Sarah Z. Dungan, Lauren E. Dares, and Claryana Arau´jo-Wang. Biology and Conservation of the Taiwanese Humpback Dolphin, Sousa chinensis taiwanensis. pp. 91–118.

Bingyao Chen, Xinrong Xu, Thomas A. Jefferson, Paula A. Olson, Qiurong Qin, Hongke Zhang, Liwen He, and Guang Yang. Conservation Status of the Indo-Pacific Humpback Dolphin (Sousa chinensis) in the Northern Beibu Gulf, China. pp. 119–140.

Gianna Minton, Anna Norliza Zulkifli Poh, Cindy Peter, Lindsay Porter, and Danielle Kreb. Indo-Pacific Humpback Dolphins in Borneo: A Review of Current Knowledge with Emphasis on Sarawak. pp. 141–156.

Guido J. Parra and Daniele Cagnazzi. Conservation Status of the Australian Humpback Dolphin (Sousa sahulensis) Using the IUCN Red List Criteria. pp. 157–192.

Daniella M. Hanf, Tim Hunt, and Guido J. Parra. Humpback Dolphins of Western Australia: A Review of Current Knowledge and Recommendations for Future Management. pp. 193–218.

Isabel Beasley, Maria Jedensjö, Gede Mahendra Wijaya, Jim Anamiato, Benjamin Kahn, and Danielle Kreb. Observations on Australian Humpback Dolphins (Sousa sahulensis) in Waters of the Pacific Islands and New Guinea. pp. 219–272.

Alexander M. Brown, Lars Bejder, Guido J. Parra, Daniele Cagnazzi, Tim Hunt, Jennifer L. Smith, and Simon J. Allen. Sexual Dimorphism and Geographic Variation in Dorsal Fin Features of Australian Humpback Dolphins, Sousa sahulensis. pp. 273–314.

Volume 74, 2016.

J. Salinger, A.J. Hobday, R.J. Matear, T.J. O'Kane, J.S. Risbey, P. Dunstan, J.P. Eveson, E.A. Fulton, M. Feng, É.E. Plagányi, E.S. Poloczanska, A.G. Marshall, and P.A. Thompson. Decadal-Scale Forecasting of Climate Drivers for Marine Applications. pp. 1–68.

S.A. Foo and M. Byrne. Acclimatization and Adaptive Capacity of Marine Species in a Changing Ocean. pp. 69–116.

N.D. Gallo and L.A. Levin. Fish Ecology and Evolution in the World's Oxygen Minimum Zones and Implications of Ocean Deoxygenation. pp. 117–198.

R.J. Olson, J.W. Young, F. Ménard, M. Potier, V. Allain, N. Goñi, J.M. Logan, and F. Galván-Magaña. Bioenergetics, Trophic Ecology, and Niche Separation of Tunas. pp. 199–344.

Volume 75, 2016.

G. Notarbartolo di Sciara. Marine Mammals in the Mediterranean Sea: An Overview. pp. 1–36.

L. Rendell and A. Frantzis. Mediterranean Sperm Whales, *Physeter macrocephalus*: The Precarious State of a Lost Tribe. pp. 37–74.

G. Notarbartolo di Sciara, M. Castellote, J.-N. Druon, and S. Panigada. Fin Whales, *Balaenoptera physalus*: At Home in a Changing Mediterranean Sea?. pp. 75–102.

M. Podestà, A. Azzellino, A. Cañadas, A. Frantzis, A. Moulins, M. Rosso, P. Tepsich, and C. Lanfredi. Cuvier's Beaked Whale, *Ziphius cavirostris*, Distribution and Occurrence in the Mediterranean Sea: High-Use Areas and Conservation Threats. pp. 103–140.

R. Esteban, P. Verborgh, P. Gauffier, D. Alarcón, J.M. Salazar-Sierra, J. Giménez, A.D. Foote, and R. de Stephanis. Conservation Status of Killer Whales, *Orcinus orca*, in the Strait of Gibraltar. pp. 141–172.

P. Verborgh, P. Gauffier, R. Esteban, J. Giménez, A. Cañadas, J.M. Salazar-Sierra, and R. de Stephanis. Conservation Status of Long-Finned Pilot Whales, *Globicephala melas*, in the Mediterranean Sea. pp. 173–204.

A. Azzellino, S. Airoldi, S. Gaspari, C. Lanfredi, A. Moulins, M. Podestà, M. Rosso, and P. Tepsich. Risso's Dolphin, *Grampus griseus*, in the Western Ligurian Sea: Trends in Population Size and Habitat Use. pp. 205–232.

D. Kerem, O. Goffman, M. Elasar, N. Hadar, A. Scheinin, and T. Lewis. The Rough-Toothed Dolphin, *Steno bredanensis*, in the Eastern Mediterranean Sea: A Relict Population?. pp. 233–258.

J. Gonzalvo, G. Lauriano, P.S. Hammond, K.A. Viaud-Martinez, M.C. Fossi, A. Natoli, and L. Marsili. The Gulf of Ambracia's Common Bottlenose Dolphins, *Tursiops truncatus*: A Highly Dense and yet Threatened Population. pp. 259–296.

G. Bearzi, S. Bonizzoni, N.L. Santostasi, N.B. Furey, L. Eddy, V.D. Valavanis, and O. Gimenez. Dolphins in a Scaled-Down Mediterranean: The Gulf of Corinth's Odontocetes. pp. 297–332.

M.C. Fontaine. Harbour Porpoises, *Phocoena phocoena*, in the Mediterranean Sea and Adjacent Regions: Biogeographic Relicts of the Last Glacial Period. pp. 333–358.

G. Notarbartolo di Sciara and S. Kotomatas. Are Mediterranean Monk Seals, *Monachus monachus*, Being Left to Save Themselves from Extinction?. pp. 359–386.

T. Scovazzi. The International Legal Framework for Marine Mammal Conservation in the Mediterranean Sea. pp. 387–416.

Volume 76, 2017.

K.S. Meyer. Islands in a Sea of Mud: Insights From Terrestrial Island Theory for Community Assembly on Insular Marine Substrata. pp. 1–40.

E.M. Montgomery, J.-F. Hamel, and A. Mercier. Patterns and Drivers of Egg Pigment Intensity and Colour Diversity in the Ocean: A Meta-Analysis of Phylum Echinodermata. pp. 41–104.

F. Espinosa and G.A. Rivera-Ingraham. Biological Conservation of Giant Limpets: The Implications of Large Size. pp. 105–156.

L. Yebra, T. Kobari, A.R. Sastri, F. Gusmão, and S. Hernández-León. Advances in Biochemical Indices of Zooplankton Production. pp. 157–240.

Volume 77, 2017.

Dayv Lowry and Shawn E. Larson. Introduction to Northeast Pacific Shark Biology, Ecology, and Conservation. pp. 1–8.

David A. Ebert, Jennifer S. Bigman, and Julia M. Lawson. Biodiversity, Life History, and Conservation of Northeastern Pacific Chondrichthyans pp. 9–78.

Shawn E. Larson, Toby S. Daly-Engel, and Nicole M. Phillips. Review of Current Conservation Genetic Analyses of Northeast Pacific Sharks. pp. 79–110.

Joseph J. Bizzarro, Aaron B. Carlisle, Wade D. Smith, and Enric Cortés. Diet Composition and Trophic Ecology of Northeast Pacific Ocean Sharks. pp. 111–148.

Jonathan C.P. Reum, Gregory D. Williams, and Chris J. Harvey. Stable Isotope Applications for Understanding Shark Ecology in the Northeast Pacific Ocean. pp. 149–178.

Mary E. Matta, Cindy A. Tribuzio, David A. Ebert, Kenneth J. Goldman, and Christopher M. Gburski. Age and Growth of Elasmobranchs and Applications to Fisheries Management and Conservation in the Northeast Pacific Ocean. pp. 179–220.

Volume 78, 2017.

Shawn E. Larson and Dayv Lowry. Introduction to Northeast Pacific Shark Biology, Research, and Conservation, Part B. pp. 1–8.

Jackie King, Gordon A. McFarlane, Vladlena Gertseva, Jason Gasper, Sean Matson, and Cindy A. Tribuzio. Shark Interactions With Directed and Incidental Fisheries in the Northeast Pacific Ocean: Historic and Current Encounters, and Challenges for Shark Conservation. pp. 9–44.

Dovi Kacev, Timothy J. Sippel, Michael J. Kinney, Sebastián A. Pardo, and Christopher G. Mull. An Introduction to Modelling Abundance and Life History Parameters in Shark Populations. pp. 45–88.

Michael Grassmann, Bryan McNeil, and Jim Wharton. Sharks in Captivity: The Role of Husbandry, Breeding, Education, and Citizen Science in Shark Conservation. pp. 89–120.

Peter A. Mieras, Chris Harvey-Clark, Michael Bear, Gina Hodgin, and Boone Hodgin. The Economy of Shark Conservation in the Northeast Pacific: The Role of Ecotourism and Citizen Science. pp. 121–154.

Dayv Lowry. Conclusions: The Future of Shark Management and Conservation in the Northeast Pacific Ocean. pp. 155–164.

Volume 79, 2018.

Patricia L.M. Lee, Gail Schofield, Rebecca I. Haughey, Antonios D. Mazaris, and Graeme C. Hays. A Review of Patterns of Multiple Paternity Across Sea Turtle Rookeries. pp. 1–32.

Johanne Vad, Georgios Kazanidis, Lea-Anne Henry, Daniel O.B. Jones, Ole S. Tendal, Sabine Christiansen, Theodore B. Henry, and J. Murray Roberts. Potential Impacts of Offshore Oil and Gas Activities on Deep-Sea Sponges and the Habitats They Form. pp. 33–60.

Gianmarco Ingrosso, Marco Abbiati, Fabio Badalamenti, Giorgio Bavestrello, Genuario Belmonte, Rita Cannas, Lisandro Benedetti-Cecchi, Marco Bertolino, Stanislao Bevilacqua, Carlo Nike Bianchi, Marzia Bo, Elisa Boscari, Frine Cardone, Riccardo Cattaneo-Vietti, Alessandro Cau, Carlo Cerrano, Renato Chemello, Giovanni Chimienti, Leonardo Congiu, Giuseppe Corriero, Federica Costantini, Francesco De Leo, Luigia Donnarumma, Annalisa Falace, Simonetta Fraschetti, Adriana Giangrande, Maria Flavia Gravina, Giuseppe Guarnieri, Francesco Mastrototaro, Marco Milazzo, Carla Morri, Luigi Musco, Laura Pezzolesi, Stefano Piraino, Fiorella Prada, Massimo Ponti, Fabio Rindi, Giovanni Fulvio Russo, Roberto Sandulli, Adriana Villamor, Lorenzo Zane, and Ferdinando Boero. Mediterranean Bioconstructions Along the Italian Coast. pp. 61–136.

Alex D. Rogers. The Biology of Seamounts: 25 Years on. pp. 137–224.

Volume 80, 2018.
Michael H. Schleyer and Sean N. Porter. Drivers of Soft and Stony Coral Community Distribution on the High-Latitude Coral Reefs of South Africa. pp. 1–56.
Sara González-Delgado and José Carlos Hernández. The Importance of Natural Acidified Systems in the Study of Ocean Acidification: What Have We Learned? pp. 57–100.

Volume 81, 2018.
Melissa C. Cook, Adam May, Lucas Kohl, Geert Van Biesen, Christopher C. Parrish, and Penny L. Morrill. The Potential Impact of Hydrocarbons on Mussels in Port au Port Bay, Newfoundland pp. 1–22.
Yunwen Tao, Bing Chen, Baiyu (Helen) Zhang, Zhiwen (Joy) Zhu, and Qinhong Cai. Occurrence, Impact, Analysis and Treatment of Metformin and Guanylurea in Coastal Aquatic Environments of Canada, USA and Europe pp. 23–58.
Gong Zhang, Chun Yang, Mariam Serhan, Graeme Koivu, Zeyu Yang, Bruce Hollebone, Patrick Lambert, and Carl E. Brown. Characterization of Nitrogen-Containing Polycyclic Aromatic Heterocycles in Crude Oils and Refined Petroleum Products. pp. 59–96.
Arpana Rani Datta, Qiao Kang, Bing Chen, and Xudong Ye. Fate and Transport Modelling of Emerging Pollutants from Watersheds to Oceans: A Review. pp. 97–128.
Meijin Du, Wenwen Gu, Xixi Li, Fuqiang Fan, and Yu Li. Modification of Hexachlorobenzene to Molecules with Lower Long-Range Transport Potentials Using 3D-QSAR Models with a Full Factor Experimental Design pp. 129–166.
Weiyun Lin, Xixi Li, Min Yang, Kenneth Lee, Bing Chen, and Baiyu (Helen) Zhang. Brominated Flame Retardants, Microplastics, and Biocides in the Marine Environment: Recent Updates of Occurrence, Analysis, and Impacts. pp. 167–212.
Xixi Li, Zhenhua Chu, Jiawen Yang, Minghao Li, Meijin Du, Xiaohui Zhao, Zhiwen (Joy) Zhu, and Yu Li. Synthetic Musks: A Class of Commercial Fragrance Additives in Personal Care Products (PCPs) Causing Concern as Emerging Contaminants. pp. 213–280.

Volume 82, 2019.
Junaid S. Khan, Jennifer F. Provencher, Mark R. Forbes, Mark L. Mallory, Camille Lebarbenchon, and Karen D. McCoy. Parasites of seabirds: A survey of effects and ecological implications. pp. 1–50.

Charlotte Havermans, Holger Auel, Wilhelm Hagen, Christoph Held, Natalie S. Ensor, and Geraint A. Tarling. Predatory zooplankton on the move: *Themisto* amphipods in high-latitude marine pelagic food webs. pp. 51–92.

Angel Borja, Guillem Chust, and Iñigo Muxika. Forever young: The successful story of a marine biotic index. pp. 93–128.

Ferdinando Boero, Francesco De Leo, Simonetta Fraschetti, and Gianmarco Ingrosso. The Cells of Ecosystem Functioning: Towards a holistic vision of marine space. pp. 129–154.

Volume 83, 2019.

Dayv Lowry and Shawn Larson. Introduction: The sharks of Pacific Mexico and their conservation: Why should we care?. pp. 1–10.

Luz Erandi Saldaña-Ruiz, Emiliano García-Rodríguez, Juan Carlos Pérez-Jiménez, Javier Tovar-Ávila, and Emmanuel Rivera-Téllez. Biodiversity and Conservation of Sharks in Pacific Mexico. pp. 11–60.

Felipe Galván-Magaña, José Leonardo Castillo-Geniz, Mauricio Hoyos-Padilla, James Ketchum, A. Peter Klimley, Sergio Ramírez-Amaro, Yassir Eden Torres-Rojas, and Javier Tovar-Avila. Shark ecology, the role of the apex predator and current conservation status. pp. 61–114.

Jonathan Sandoval-Castillo. Conservation genetics of elasmobranchs of the Mexican Pacific Coast, trends and perspectives. pp. 115–158.

Volume 84, 2019.

Brian Morton and Fabrizio Marcondes Machado. Predatory marine bivalves: A review. pp. 1–98.

Volume 85, 2020.

James T. Ketchum, Mauricio Hoyos-Padilla, Alejandro Aldana-Moreno, Kathryn Ayres, Felipe Galván-Magaña, Alex Hearn, Frida Lara-Lizardi, Gador Muntaner-López, Miquel Grau, Abel Trejo-Ram´ırez, Darren A. Whitehead, and A. Peter Klimley. Shark movement patterns in the Mexican Pacific: A conservation and management perspective. pp. 1–38.

Oscar Sosa-Nishizaki, Emiliano García-Rodríguez, Christian D. Morales-Portillo, Juan C. Pérez-Jiménez, M. del Carmen Rodríguez-Medrano, Joseph J. Bizzarro, and José Leonardo Castillo-Géniz. Fisheries interactions and the challenges for target and nontargeted take on shark conservation in the Mexican Pacific. pp. 39–70.

A.M. Cisneros-Montemayor, E.E. Becerril-García, O. Berdeja-Zavala, and A. Ayala-Bocos. Shark ecotourism in Mexico: Scientific research, conservation, and contribution to a Blue Economy. pp. 71–92.
Oscar Sosa-Nishizaki, Felipe Galván-Magaña, Shawn E. Larson, and Dayv Lowry. Conclusions: Do we eat them or watch them, or both? Challenges for conservation of sharks in Mexico and the NEP. pp. 93–102.

Volume 86, 2020.
David J. Vance and Peter C. Rothlisberg. The biology and ecology of the banana prawns: *Penaeus merguiensis* de Man and *P. indicus* H. Milne Edwards. pp. 1–140.
Jeff A. Eble, Toby S. Daly-Engel, Joseph D. DiBattista, Adam Koziol, and Michelle R. Gaither. Marine environmental DNA: Approaches, applications, and opportunities. pp. 141–170.

Volume 87, 2020.
Bernhard M. Riegl and Peter W. Glynn. Population dynamics of the reef crisis: Consequences of the growing human population. pp. 1–30.
Chiara Pisapia, Peter J. Edmunds, Holly V. Moeller, Bernhard M. Riegl, Mike McWilliam, Christopher D. Wells, and Morgan S. Pratchett. Projected shifts in coral size structure in the Anthropocene. pp. 31–60.
Alex E. Mercado-Molina, Alberto M. Sabat, and Edwin A. Hernández-Delgado. Population dynamics of diseased corals: Effects of a Shut Down Reaction outbreak in Puerto Rican *Acropora cervicornis*. pp. 61–82.
Elizabeth A. Goergen, Kathleen Semon Lunz, and David S. Gilliam. Spatial and temporal differences in *Acropora cervicornis* colony size and health. pp. 83–114.
Rian Prasetia, Zi Wei Lim, Aaron Teo, Tom Shlesinger, Yossi Loya, and Peter A. Todd. Population dynamics and growth rates of free-living mushroom corals (Scleractinia: Fungiidae) in the sediment-stressed reefs of Singapore. pp. 115–140.
Joshua S. Feingold and Brandon Brulé. Population fluctuations of the fungiid coral *Cycloseris curvata*, Galápagos Islands, Ecuador. pp. 141–166.
Stuart A. Sandin, Clinton B. Edwards, Nicole E. Pedersen, Vid Petrovic, Gaia Pavoni, Esmeralda Alcantar, Kendall S. Chancellor, Michael D. Fox, Brenna Stallings, Christopher J. Sullivan, Randi D. Rotjan, Federico Ponchio, and Brian J. Zgliczynski. Considering the rates of growth in two taxa of coral across Pacific islands. pp. 167–192.

Geórgenes Cavalcante, Filipe Vieira, Jonas Mortensen, Radhouane Ben-Hamadou, Pedro Range, Elizabeth A. Goergen, Edmo Campos, and Bernhard M. Riegl. Biophysical model of coral population connectivity in the Arabian/Persian Gulf. pp. 193–222.

S.A. Matthews, C. Mellin, and Morgan S. Pratchett. Larval connectivity and water quality explain spatial distribution of crown-of-thorns starfish outbreaks across the Great Barrier Reef. pp. 223–258.

S.A. Matthews, K. Shoemaker, Morgan S. Pratchett, and C. Mellin. COTSMod: A spatially explicit metacommunity model of outbreaks of crown-of-thorns starfish and coral recovery. pp. 259–290.

Tim R. McClanahan. Coral community life histories and population dynamics driven by seascape bathymetry and temperature variability. pp. 291–330.

William F. Precht, Richard B. Aronson, Toby A. Gardner, Jennifer A. Gill, Julie P. Hawkins, Edwin A. Hernández-Delgado, Walter C. Jaap, Tim R. McClanahan, Melanie D. McField, Thaddeus J.T. Murdoch, Maggy M. Nugues, Callum M. Roberts, Christiane K. Schelten, Andrew R. Watkinson, and Isabelle M. Côté. The timing and causality of ecological shifts on Caribbean reefs. pp. 331–360.

Howard R. Lasker, Lorenzo Bramanti, Georgios Tsounis, and Peter J. Edmunds. The rise of octocoral forests on Caribbean reefs. pp. 361–410.

Sascha C.C. Steiner, Priscilla Martínez, Fernando Rivera, Matthew Johnston, and Bernhard M. Riegl. Octocoral populations and connectivity in continental Ecuador and Galápagos, Eastern Pacific. pp. 411–442.

Peter W. Glynn, Renata Alitto, Joshua Dominguez, Ana B. Christensen, Phillip Gillette, Nicolas Martinez, Bernhard M. Riegl, and Kyle Dettloff. A tropical eastern Pacific invasive brittle star species (Echinodermata: Ophiuroidea) reaches southeastern Florida. pp. 443–472.

Volume 88, 2021.

Pat Hutchings. Potential loss of biodiversity and the critical importance of taxonomy—An Australian perspective. pp. 3–16.

Ferdinando Boero. Mission possible: Holistic approaches can heal marine wounds. pp. 19–38.

Iraide Artetxe-Arrate, Igaratza Fraile, Francis Marsac, Jessica H. Farley, Naiara Rodriguez-Ezpeleta, Campbell R. Davies, Naomi P. Clear, Peter Grewe, and Hilario Murua. A review of the fisheries, life history and stock structure of tropical tuna (skipjack *Katsuwonus pelamis*, yellowfin *Thunnus albacares* and bigeye *Thunnus obesus*) in the Indian Ocean. pp. 39–90.

Yan Ji and Kunshan Gao. Effects of climate change factors on marine macroalgae: A review. pp. 91–136.

Brian Morton and Fabrizio Marcondes Machado. The origins, relationships, evolution and conservation of the weirdest marine bivalves: The watering pot shells. A review. pp. 139–220.

Volume 89, 2021.

Stanislao Bevilacqua, Laura Airoldi, Enric Ballesteros, Lisandro Benedetti-Cecchi, Ferdinando Boero, Fabio Bulleri, Emma Cebrian, Carlo Cerrano, Joachim Claudet, Francesco Colloca, Martina Coppari, Antonio Di Franco, Simonetta Fraschetti, Joaquim Garrabou, Giuseppe Guarnieri, Cristiana Guerranti, Paolo Guidetti, Benjamin S. Halpern, Stelios Katsanevakis, Maria Cristina Mangano, Fiorenza Micheli, Marco Milazzo, Antonio Pusceddu, Monia Renzi, Gil Rilov, Gianluca Sarà, and Antonio Terlizzi. Mediterranean rocky reefs in the Anthropocene: Present status and future concerns. pp. 1–52.

Johanne Vad, Kelsey Archer Barnhill, Georgios Kazanidis, and J. Murray Roberts. Human impacts on deep-sea sponge grounds: Applying environmental omics to monitoring. pp. 53–78.

Angel Borja and Michael Elliott. From an economic crisis to a pandemic crisis: The need for accurate marine monitoring data to take informed management decisions. pp. 79–114.

Volume 90, 2021.

Shawn Larson, Dayv Lowry, Nicholas K. Dulvy, Jim Wharton, Felipe Galván-Magaña, Abraham B. Sianipar, Christopher G. Lowe, and Erin Meyer. Current and future considerations for shark conservation in the Northeast and Eastern Central Pacific Ocean. pp. 1–50.

Ferdinando Boero. The future ocean we want. pp. 51–64.

Volume 91, 2022.

Jean-François Hamel, Igor Eeckhaut, Chantal Conand, Jiamin Sun, Guillaume Caulier, and Annie Mercier. Global knowledge on the commercial sea cucumber *Holothuria scabra*. pp. 1–286.

Volume 92, 2022.

Monica Montefalcone, Alice Oprandi, Annalisa Azzola, Carla Morri, and Carlo Nike Bianchi. Serpulid reefs and their role in aquatic ecosystems: A global review. pp. 1–54.

Matthew R. Nitschke, Sabrina L. Rosset, Clinton A. Oakley, Stephanie G. Gardner, Emma F. Camp, David J. Suggett, and Simon K. Davy. The diversity and ecology of Symbiodiniaceae: A traits-based review. pp. 55–128.

David Rodríguez-Rodríguez and Javier Martínez-Vega. Ecological effectiveness of marine protected areas across the globe in the scientific literature. pp. 129–154.

Volume 93, 2022.

Fabian Ritter. Opinion: Marine mammal conservation in the 21st century: A plea for a paradigm shift towards mindful conservation. pp. 1–22.

Alex D. Rogers, Ward Appeltans, Jorge Assis, Lisa T. Ballance, Philippe Cury, Carlos Duarte, Fabio Favoretto, Lisa A. Hynes, Joy A. Kumagai, Catherine E. Lovelock, Patricia Miloslavich, Aidin Niamir, David Obura, Bethan C. O'Leary, Eva Ramirez-Llodra, Gabriel Reygondeau, Callum Roberts, Yvonne Sadovy, Oliver Steeds, Tracey Sutton, Derek P. Tittensor, Enriqueta Velarde, Lucy Woodall, and Octavio Aburto-Oropeza. Discovering marine biodiversity in the 21st century. pp. 23–116.

Kailei Zhu and Jiayu Bai. Review study on governance and international law for coastal and marine ecosystems in response to climate change: Social science perspective. pp. 117–146.

Volume 94, 2023.

Jesse van der Grient, Simon Morley, Alexander Arkhipkin, James Bates, Alastair Baylis, Paul Brewin, Michael Harte, J. Wilson White, and Paul Brickle. The Falkland Islands marine ecosystem: A review of the seasonal dynamics and trophic interactions across the food web. pp. 1–68.

Tamara A. Shiganova, Andrei M. Kamakin, Larisa A. Pautova, Alexander S. Kazmin, Aboulghasem Roohi, and Henri J. Dumont. An impact of non-native species invasions on the Caspian Sea biota. pp. 69–158.

Luís Gabriel A. Barboza, Sara Couto Lourenço, Alexandre Aleluia, Natália Carneiro Lacerda dos Santos, Minrui Huang, Jun Wang, and Lúcia Guilhermino. A global synthesis of microplastic contamination in wild fish species: Challenges for conservation, implications for sustainability of wild fish stocks and future directions. pp. 159–200.

Nikolina Rako-Gospić and Marta Picciulin. Addressing underwater noise: Joint efforts and progress on its global governance. pp. 201–232.

Volume 95, 2023.
Julie A. Lively and Jonathan McKenzie. Discards and bycatch: A review of wasted fishing. pp. 1–26.
James J. Bell, Francesca Strano, Manon Broadribb, Gabriela Wood, Ben Harris, Anna Carolina Resende, Emma Novak, and Valerio Micaroni. Sponge functional roles in a changing world. pp. 27–90.
Ferdinando Boero and Joachim Mergeay. Darwin's feathers: Ecoevolutionary biology, predictions and policy. pp. 91–112.
Mauvis Gore, Ewan Camplisson, and Rupert Ormond. The biology and ecology of the basking shark: A review. pp. 113–257.

Volume 96, 2024.
Ayaka T. Matsuda, Takashi F. Matsuishi, Fumika Noto, Masao Amano, Yuko Tajima, and Tadasu K. Yamada. Notes on stomach contents of pygmy and dwarf sperm whales (*Kogia* spp.) from around Japan. pp. 1–24.
Shin Nishida, Atsushi Uchimura, Yuko Tajima, and Tadasu K. Yamada. Comparative analysis of the genetic structures of *Kogia* spp. populations in the western North Pacific. pp. 25–37.
Akira Shiozaki, Shotaro Nakagun, Yuko Tajima, and Masao Amano. A first record of digenean parasites of the dwarf sperm whale *Kogia sima* with morphological and molecular information. pp. 39–61.
Névia Lamas, Pablo Covelo, Alfredo López, Uxía Vázquez, and Nuria Alemañ. A histological study of the facial hair follicles in the pygmy sperm whale (*Kogia breviceps*). pp. 63–83.
Stephanie Plön, Peter B. Best, Pádraig Duignan, Shane D. Lavery, Ric T.F. Bernard, Koen Van Waerebeek, and C. Scott Baker. Population structure of pygmy (*Kogia breviceps*) and dwarf (*Kogia sima*) sperm whales in the Southern Hemisphere may reflect foraging ecology and dispersal patterns. pp. 85–114.

Volume 97, 2024.
Shawna A. Foo, Pauline M. Ross, and Maria Byrne. The 2024 roadmap for understanding marine species' resilience in a changing ocean. pp. 1–10.
Shawna A. Foo and Maria Byrne. Reprint: Acclimatization and Adaptive Capacity of Marine Species in a Changing Ocean. pp. 11–58.
José Carlos Hernández, Sara González-Delgado, M. Aliende-Hernández, B. Alfonso, A. Rufino-Navarro, and C.A. Hernández. Natural acidified marine systems: Lessons and predictions. pp. 59–78.

Sara González-Delgado and José Carlos Hernández. Reprint: The Importance of Natural Acidified Systems in the Study of Ocean Acidification: What Have We Learned?. pp. 79–122.

Patricia L.M. Lee and Graeme C. Hays. A roadmap for multiple paternity research with sea turtles. pp. 123–134.

Patricia L.M. Lee, Gail Schofield, Rebecca I. Haughey, Antonios D. Mazaris, and Graeme C. Hays. Reprint: A Review of Patterns of Multiple Paternity Across Sea Turtle Rookeries. pp. 135–166.

Ferdinando Boero. A roadmap to knowledge-based maritime spatial planning. pp. 167–190.

Ferdinando Boero, Francesco De Leo, Simonetta Fraschetti, and Gianmarco Ingrosso. Reprint: The Cells of Ecosystem Functioning: Towards a holistic vision of marine space. pp. 191–216.

Volume 98, 2024.

Götz B. Reinicke, Sabine Holst, André C. Morandini, Ilka Sötje, Ilka Straehler-Pohl, Amanda A. Wiesenthal, and Hjalmar Thiel. Max Egon Thiel's monographs on Scyphozoa (Cnidaria) and a left-behind typescript on the Rhizostomeae. pp. 1–60.

André C. Morandini. Morphology of Rhizostomeae jellyfishes: What is known and what we advanced since the 1970s. pp. 61–98.

Sabine Holst, Gisele R. Tiseo, Nicolas Djeghri, and Ilka Sötje. Approaches and findings in histological and micromorphological research on Rhizostomeae. pp. 99–192.

Agustín Schiariti, Sabine Holst, Gisele R. Tiseo, Hiroshi Miyake, and André C. Morandini. Life cycles and reproduction of Rhizostomeae. pp. 193–254.

Renato M. Nagata, Isabella D'Ambra, Chiara Lauritano, Guilherme M. von Montfort, Nicolas Djeghri, Mayara A. Jordano, Sean P. Colin, John H. Costello, and Valentina Leoni. Physiology and functional biology of Rhizostomeae jellyfish. pp. 255–360.

Edgar Gamero-Mora, Jonathan W. Lawley, Maximiliano M. Maronna, Sérgio N. Stampar, Adriana Muhlia-Almazan, and André C. Morandini. Morphological and molecular data in the study of the evolution, population genetics and taxonomy of Rhizostomeae. pp. 361–396.

Delphine Thibault, Zafrir Kuplik, Laura Prieto, Angelica Enrique-Navarro, Michael Brown, Shin Uye, Tom Doyle, Kylie Pitt, William Fitt, and Mark Gibbons. Ecology of Rhizostomeae. pp. 397–510.

Lucas Brotz, Dror L. Angel, Isabella D'Ambra, Angélica Enrique-Navarro, Chiara Lauritano, Delphine Thibault, and Laura Prieto. Rhizostomes as a resource: The expanding exploitation of jellyfish by humans. pp. 511–548.

CHAPTER ONE

Habitat suitability, occurrence, and behavior of dwarf sperm whales (*Kogia sima*) off St. Vincent and the Grenadines, Eastern Caribbean

Jeremy J. Kiszka*, Guilherme Maricato, and Michelle Caputo

Institute of Environment, Department of Biological Sciences, Florida International University, North Miami, United States
*Corresponding author. e-mail address: jkiszka@fiu.edu

Contents

1. Introduction	2
2. Materials and methods	4
2.1 Data collection	4
2.2 Data analysis	5
3. Results	8
3.1 Occurrence and behavior	8
3.2 Habitat suitability	8
3.3 Photo-identification	9
4. Discussion	11
Acknowledgments	15
Appendix A. Supporting information	16
References	16

Abstract

The genus *Kogia* includes two species that are some of the least known cetacean species around the globe. Here, we investigated the occurrence, behavior, and habitat suitability of dwarf sperm whales (*K. sima*) off St. Vincent and the Grenadines (Eastern Caribbean). Small boat dedicated surveys were conducted during May and June of both 2022 and 2023 along the south and west coast of the island of St. Vincent. A total of 2260 km was surveyed and 33 sightings of dwarf sperm whale were recorded, which was also the most frequently sighted cetacean species (37.5% of all cetacean sightings). Group size varied from 1 to 20 individuals (mean = 2.08, SD = 3.23). Traveling and breaching were the most commonly recorded behavioral categories and occurred at an equal proportion (28.6%). The distribution of dwarf sperm whales was restricted to the south and southwest portion of St. Vincent in depths ranging from 95 to 1104 m (mean = 650 m).

Habitat suitability (in relation to depth and slope) was investigated using an ensemble model using three algorithms (GLM, GAM, and MaxEnt). The model revealed that slope, and to a lesser extent depth, were important in explaining the habitat suitability of dwarf sperm whales. This preliminary research highlights the existence of a globally important area for dwarf sperm whales off St. Vincent, where encounter rates are significantly higher than in any other known island-associated habitat.

1. Introduction

Odontocetes of the genus *Kogia* include two currently recognized species: pygmy (*K. breviceps*) and dwarf sperm whales (*K. sima*), both among the least known cetacean species globally. Although the distribution of both species overlaps in tropical and subtropical regions, *K. breviceps* tends to be more common in temperate waters, whereas *K. sima* almost exclusively occurs in tropical and subtropical regions (McAlpine, 2018; Kiszka & Braulik, 2020, 2022). Dwarf sperm whales occurring in the Atlantic might constitute a third species, separated from those occurring in the Indo-Pacific region, although further evidence is required before this can be confirmed (Chivers et al., 2005; Kiszka & Braulik, 2020, 2022; Plön et al., 2023). Both species of *Kogia* are globally distributed in deep oceanic waters, particularly over shelf break and abyssal plains, where they forage on vertically migrating oceanic cephalopods (Spitz et al., 2011; Staudinger et al., 2014; Moura et al., 2016). The rarity of live sightings of both species suggest that population densities are low throughout their range (e.g., Garrison et al., 2010; Barlow, 2015). The inconspicuous behavior of dwarf and pygmy sperm whales probably contributes to the rarity of live reports (McAlpine, 2018; Plön et al., 2022; Plön & Baird, 2022). However, stranding records and passive acoustic monitoring indicate that cetaceans of the genus *Kogia* may be common in some regions, such as along the east coast of South Africa (Plön, 2004), or off the southeastern United States, where *K. breviceps* is the second most frequently stranded cetacean after the common bottlenose dolphin (*Tursiops truncatus*; Odell, 1991), and *Kogia* are frequently detected acoustically (Hodge et al., 2018), but rarely seen (Garrison et al., 2010). Because both species of *Kogia* are so cryptic and rarely observed at sea, and due to the difficulty of distinguishing *K. breviceps* from *K. sima*, there are very few abundance estimates available for either species (Palka, 2012; Garrison, 2016; Laran et al., 2017). Although both species are listed as Least Concern on the IUCN Red List of Threatened Species (Kiszka & Braulik, 2020, 2022), several threats have been identified, including noise pollution (Yong et al., 2008), whaling

(Caldwell et al., 1973; Ilangakoon, 2012; Fielding & Kiszka, 2021), harmful algal blooms (Fire et al., 2009), and fisheries interactions (Baird et al., 2021), including bycatch in offshore drift gillnets (Kiszka et al., 2021a). However, the magnitude of all these threats is poorly known.

Dwarf sperm whales have been found to regularly occur in some regions, particularly in slope and deep waters around oceanic islands, such as Abaco, Bahamas (MacLeod et al., 2004; Dunphy-Daly et al., 2008), Mayotte, the Maldives and the Seychelles in the western tropical Indian Ocean (Kiszka et al., 2010; Laran et al., 2017), and Hawai'i (Baird et al., 2013, 2021). The presence of island-associated populations has allowed researchers to unravel some aspects of the ecology and behavior of dwarf sperm whales, including their habitat preferences (particularly in relation to physiography), group dynamics (size, composition), site fidelity, and social structure (Dunphy-Daly et al., 2008; Baird et al., 2021). The existence of island-associated populations also gives an opportunity to further explore the role of environmental variables on the distribution and abundance of dwarf sperm whales, and therefore predict how this species might respond to climate change (van Weelden et al., 2021; Lettrich et al., 2023). Although describing and predicting how cetaceans are distributed is critical to support management efforts, there have been limited efforts to assess the influence of environmental predictors on the spatiotemporal distribution of many deep diving cetaceans, particularly beaked (Ziphiidae) and *Kogia* whales. The main limitation in published studies is sample size, which is usually small for such species occurring at low densities and that are challenging to detect at sea (e.g., Rogan et al., 2017; Virgili et al., 2021). However, some species distribution models have relatively strong predictive capabilities, even with small sample sizes (Phillips et al., 2006; Phillips & Dudík, 2008), and are used to predict suitable habitats based on the characteristics of the environment (e.g., Redfern et al., 2006; Fiedler et al., 2023). Physiography (depth and slope) has been successfully used to predict the distribution of cetaceans, particularly deep-diving whales (e.g., Cañadas et al., 2002; Rogan et al., 2017; Virgili et al., 2021). Here, we employed three commonly used algorithms (Generalized Linear Models, General Additive Models, and MaxEnt) to create an ensemble model to understand the influence of depth and slope on the distribution of dwarf sperm whales around St. Vincent and the Grenadines (SVG), in the Eastern Caribbean. We also document the encounter rates, group dynamics (size, composition), and some aspects of the behavior of this species during multi-species cetacean survey data conducted in May and June of both 2022 and 2023.

2. Materials and methods
2.1 Data collection

Field surveys took place in SVG, a small archipelagic nation located in the eastern Caribbean region (Fig. 1). The main island of St. Vincent is of volcanic origin and is surrounded by a steep insular slope; the waters are oligotrophic, highly stratified, and mostly influenced by the North Equatorial current. Small boat (2 boats, 8–9 m) surveys were carried out off the south and west coast of St. Vincent in May and June of both 2022 and 2023 from 8:00 to17:00 at a speed of 7–10 knots, with the initial objective of assessing the spatial distribution, encounter rates, and habitat characteristics of cetaceans. Transect location and surveys were randomly designed but could be adapted if sea conditions and wind changed (Fig. 1). At least three observers (excluding the pilot) conducted the surveys with the naked eye. Sea state was collected using the Beaufort scale throughout the day, particularly when changes occurred. Survey effort was interrupted if sea state was B > 4. Survey effort was quantified both spatially and in time using a handheld Global Positioning System. For each sighting, cetaceans were identified to species level, group size was estimated (minimum, maximum, and best estimate), the time (hh:mm:ss) and

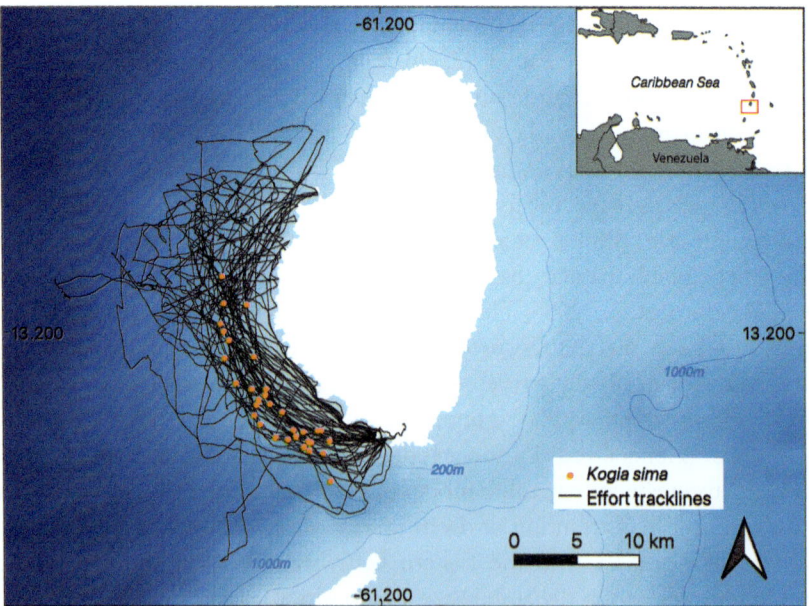

Fig. 1 Spatial distribution of *Kogia sima* sightings (dots) and survey effort (lines) off the island of St. Vincent in May-June 2022 and 2023.

geographic position were recorded. A sighting refers to both single individuals and groups, which are defined as two or more individuals and engaged in a similar behavioral activity in small cetaceans, particularly delphinids (Syme et al., 2022). As dwarf sperm whales can be widely spaced (up to about 200 m; Baird et al., 2021), individuals were considered as a group if they were within 300 m, exhibited the same behavior, orientation, and speed. The initial behavior of cetaceans encountered was also recorded and included traveling, socializing, foraging, milling, and resting. For dwarf sperm whales, only three behavioral categories were identified: traveling, milling, and resting. Resting was characterized by logging at the surface, motionless. Traveling was characterized by consistent (at least 3 successive individual/group surfacing) directional movements, whereas milling consisted in non-directional movements, with frequent changes in heading. When breaching, individuals usually exposing their full body, except the fluke. Breaching was not considered as a behavioral category per se but was categorized as such when no behavioral activity could be assigned to the group or individual observed. Group composition was recorded based on the estimated relative body length of individuals encountered and consisted of three main categories: mature adult, juvenile, and calf. If a group or individual was lost, the reason for ending the encounter was documented (sampling completed, individual/group lost).

The use of natural markings on the dorsal fin and adjacent areas of the dorsal region has previously been used to identify individual dwarf sperm whales in other regions, particularly off the island of Hawai'i (Baird et al., 2021). Photographs were also opportunistically collected using a DSLR camera (Canon 7D Mark II, 100–400 mm lens) during several encounters. This allowed us to investigate the potential use of this methodology to examine site fidelity, association patterns, and demographic parameters of this population. Photos were taken of the encounter of 20 animals and one is included as a Supplementary Figure. In addition, photographs were also used to collect information on skin condition, scaring patterns, and potential health status of individuals encountered (Van Bressem et al., 2015; Kiszka et al., 2008; Van Bressem et al., 2009, 2015). When potential lesions were identified, photographs were submitted to two independent experts (veterinarians with expertise in cetacean pathologies) for identification.

2.2 Data analysis

The influence of sea state (in Beaufort) on encounter rates of dwarf sperm whales was assessed using a one-way ANOVA. Depth and slope were the environmental variables used to assess the habitat suitability of dwarf sperm whales in SVG (see supplementary, Fig. 1), as they usually best explain the distribution of deep diving odontocetes (Cañadas et al., 2002; Rogan et al., 2017; Virgili et al., 2021). Depth was obtained from the General Bathymetric Chart of the Oceans; slope was calculated using the depth layer through the surface parameters tool in ArcGIS Pro 3.2.1. The resolution of each layer was 0.5 km. Habitat suitability modeling was performed in the R environment (R 4.3.2, RStudio 2023.12.0), using the biomod2 package (Thuiller et al., 2023). Occurrence data were filtered using the spThin package (Aiello-Lammens et al., 2015) to remove spatial autocorrelation, and 10 sets of pseudo-absences were generated using the random strategy (Guisan et al., 2017). Three algorithms were used, including presence-absence Generalized Linear Models (GLMs), Generalized Additive Models (GAMs), and the Maximum Entropy (MaxEnt), a presence-background machine learning algorithm. For model calibration, a 10 km buffer around SVG was created using the pairwise buffer tool in ArcGIS Pro 3.2.1, which was based on previous studies on the depth preferences of dwarf sperm whales around other oceanic islands: Abaco, Bahamas (400–1600 m; Dunphy-Daly et al., 2008) and Hawai'i (500–1500 m; Baird, 2005; Baird et al., 2013). Then, we divided 70% of the occurrence data for training and 30% for testing using k-fold cross-validation with $k = 5$, repeating each set of analyses 10 times. We used the Area Under the Curve (AUC) of the Receiver Operating Characteristic to evaluate each model, considering those with an evaluation greater than or equal to 0.7 for the ensemble model. The ensemble model was created from the weighted mean of the selected individual models, in which models with better performances had a greater weight. The relative importance of each variable (depth, slope) was obtained by assessing the model predictions' sensitivity to each of them, by shuffling one variable at a time, and calculating the correlation between the reference predictions and the predictions with the shuffled data. The variable importance score is represented by 1 minus the correlation, indicating that the higher the value, the greater the variable's influence on the model (Guisan et al., 2017). Following a standard protocol (Zurell et al., 2020), we present the committee average map as a measure to assess model uncertainty (Supplementary, Fig. 2). The committee average consists of the binary frequency of individual models, that is extreme values (0 and 100) mean that all individual models identified the area as unsuitable or suitable, respectively. Intermediate values (close to 50) indicate that some models identified the area as unsuitable, while others considered it suitable, that is there was no consensus, and the results for that area are uncertain.

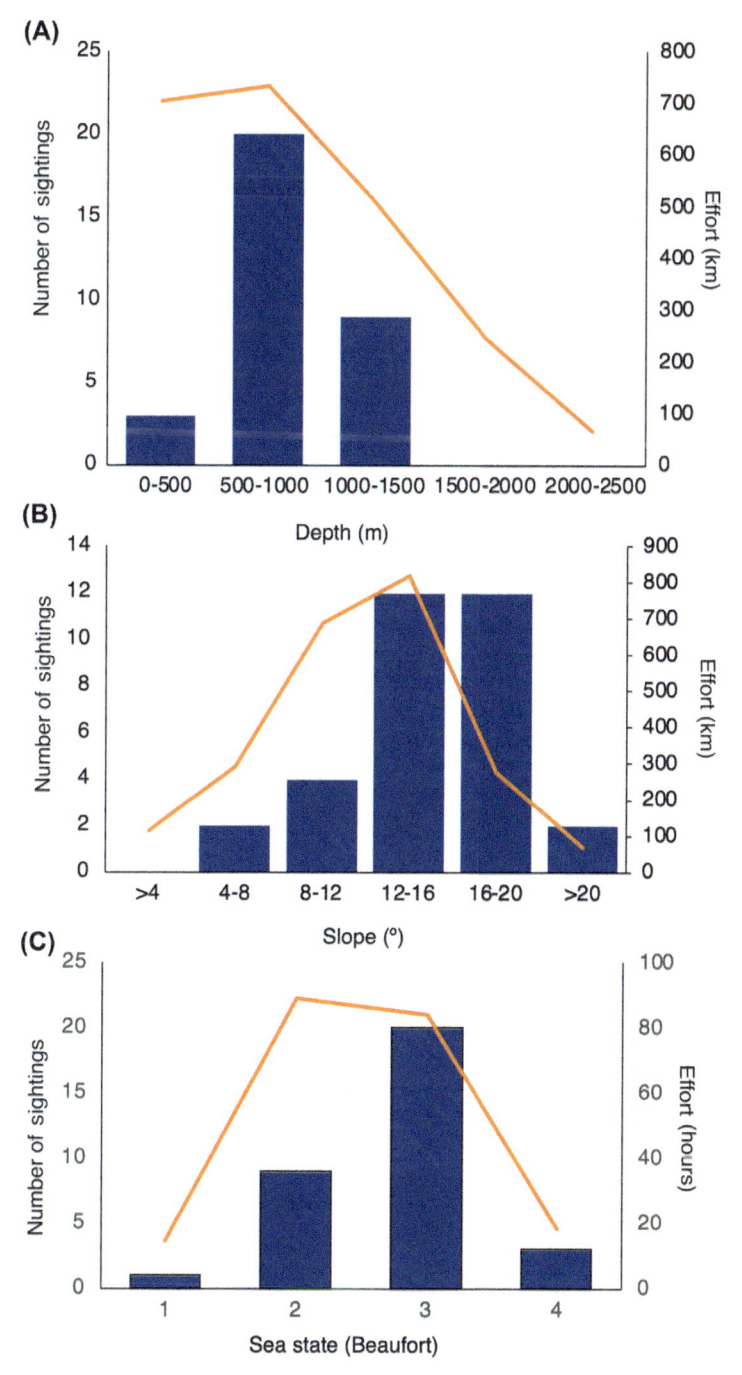

Fig. 2 Relationship between (A) effort (orange line, in km) and the number of sightings of *K. sima* (blue bars) across each depth ranges, (B) slope ranges, and (C) sea states (in Beaufort scale) off the island of St. Vincent in May-June 2022 and 2023.

3. Results

3.1 Occurrence and behavior

The waters off the south and west coasts of St. Vincent were surveyed for 206 h over 33 day and a cumulative distance of 2260 km (Fig. 1). A total of 33 sightings of dwarf sperm whales were recorded, out of a total of 88 cetacean encounters (37.5% of all cetacean sightings) throughout the study period. Group size ranged from 1 to 20 individuals (mean = 2.08; SD = 3.23), although 66.6% of encounters involved single individuals. A large group of at least 20 individuals, primarily made up of adults and a possible mother-calf pair, was observed traveling on May 15th, 2023 (see supplementary material). This group was observed in a tight group formation in the vicinity (100–150 m) of a group of 200–250 Clymene dolphins (*Stenella clymene*). The group was forming a narrow line and was lost 6 min after being found and exhibited an evasive behavior towards the research vessel. Overall, the duration of encounters was short and ranged from less than a minute, typically when the animal(s) was sighted and dive shortly after being detected, to 36 min (median = 6.1 min).

Traveling and breaching were the behavioral categories recorded at an equal proportion (28.6%). In one instance, a traveling group of 3 dwarf sperm whales was also observed breaching on two occasions (involving two individuals). Milling was observed for 11.4% of groups, while resting was observed for 2.8% of them (one sighting). For the remaining sightings, no behavioral activity was assigned, mostly due to the brevity of encounters. A total of three sightings included confirmed mother-calf pairs (9.1% of all sightings). Dwarf sperm whales were encountered at mean depth of 650 m (range = 95–1104; SD = 271; Fig. 2A), and over slopes of 12–20° (72.7%, Fig. 2B). The number of sightings per survey day varied from 0 to 4 (mean = 1.06, mode = 1, SD = 0.93), and encounter rates were not influenced by sea state ($F = 1.40$; $P = 0.251$). Most sightings occurred when Beaufort = 3 (Fig. 2C). When Beaufort was either 2 or 3, the majority (87%) of detections of dwarf sperm whales were when animals breached (estimated distance from boat from 30–400 m).

3.2 Habitat suitability

Spatial filtering retained 23 out of 33 total occurrences of the dwarf sperm whale (see supplementary, Fig. 3). After evaluating the performance of individual models, 1451 out of 1500 achieved AUC values ≥0.7 and formed the ensemble model (Fig. 3).

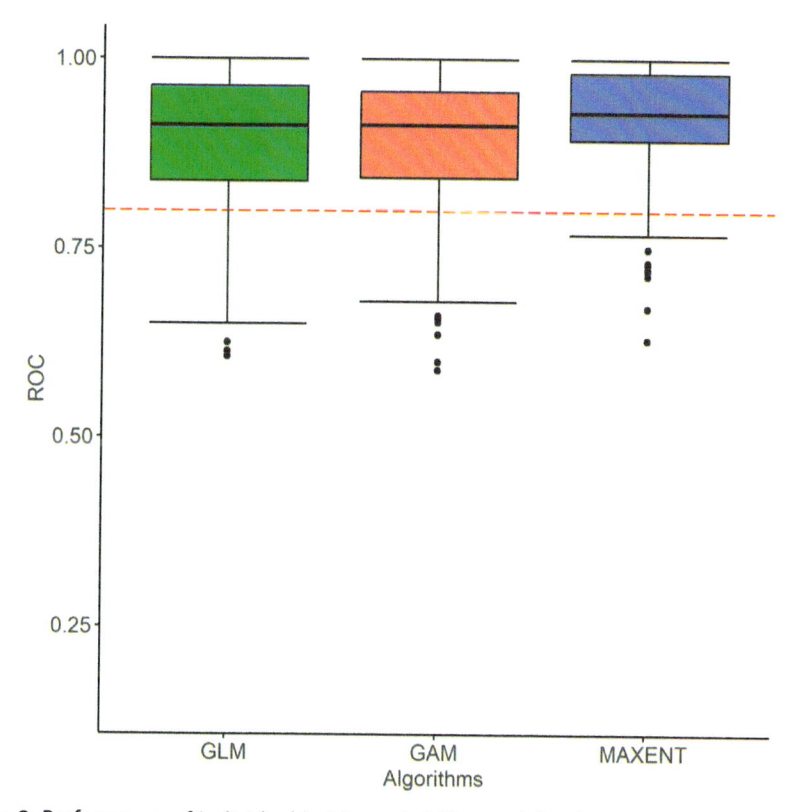

Fig. 3 Performance of individual habitat suitability models of K. sima off the island of St. Vincent. ROC, Receiver Operating Characteristic; GLM, Generalized Linear Models; GAM, Generalized Additive Models; MAXENT, Maximum Entropy.

Slope, and to a lesser extent depth, were important variables in explaining the ensemble model (slope: mean = 0.604, SD = 0.015; depth: mean = 0.377, SD = 0.028; Fig. 4). The most suitable habitats for dwarf sperm whales were located between depths ranging from 700 to 1300 m, and on steep slopes of 20–30° (Fig. 5), located along the leeward coast of SVG, particularly along the southwest coast of the main island of St. Vincent, and along the western slope off the island of Bequia, south of the main island (Fig. 6). Although sightings were not recorded off the west-north-west area off St. Vincent, the model identified it as suitable.

3.3 Photo-identification

Opportunistic photo-identification was conducted during seven sightings with dwarf sperm whales, as observed individuals would surface on

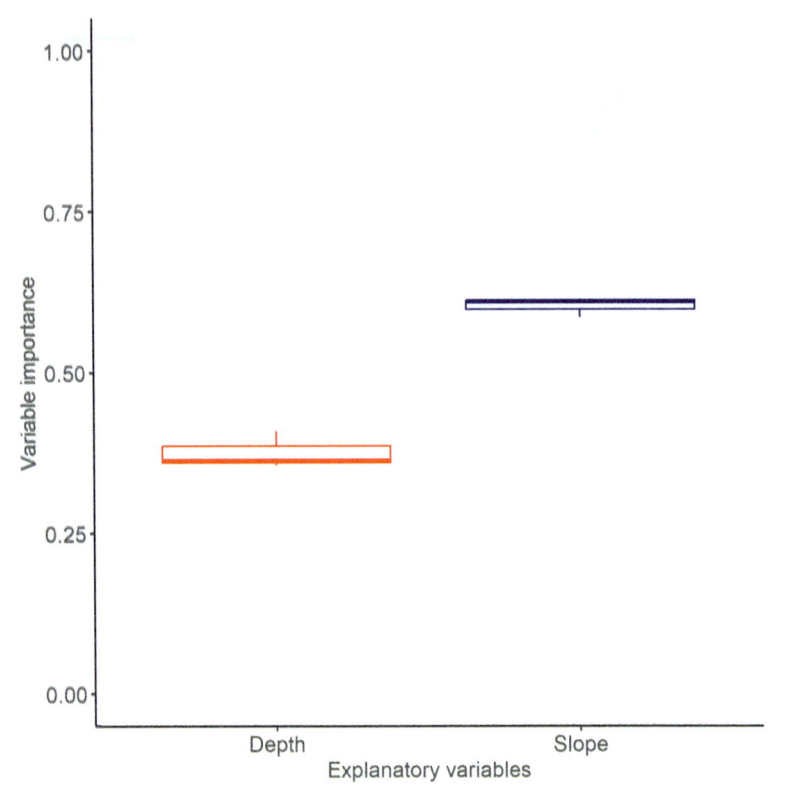

Fig. 4 Importance of explanatory variables contributing to habitat suitability models of *K. sima* off the island of St. Vincent.

multiple occasions within approximately 200 m of the research vessel (Fig. 7). With no categorizing of photograph quality and distinctiveness, a total of 22 individuals were identified, including two adult females accompanied with calves (Fig. 7B). Most photographed individuals showed distinctive patterns on their dorsal fins, including deep notches and scars on the trailing edge (Figs. 7A, 7C and 7D). No resighting of any identified individual was recorded. A total of three identified individuals ($n = 14$ photographs) exhibited skin lesions. An adult female accompanied by a calf exhibited multiple subcutaneous masses, and relative emaciation (Fig. 7A), and two other individuals exhibited dark circular lesions in the dorsal region halfway between the blowhole and the dorsal fin (Fig. 7C, individual in the foreground). One individual exhibited possible entanglement with a fishing line (Fig. 7C, individual in the background).

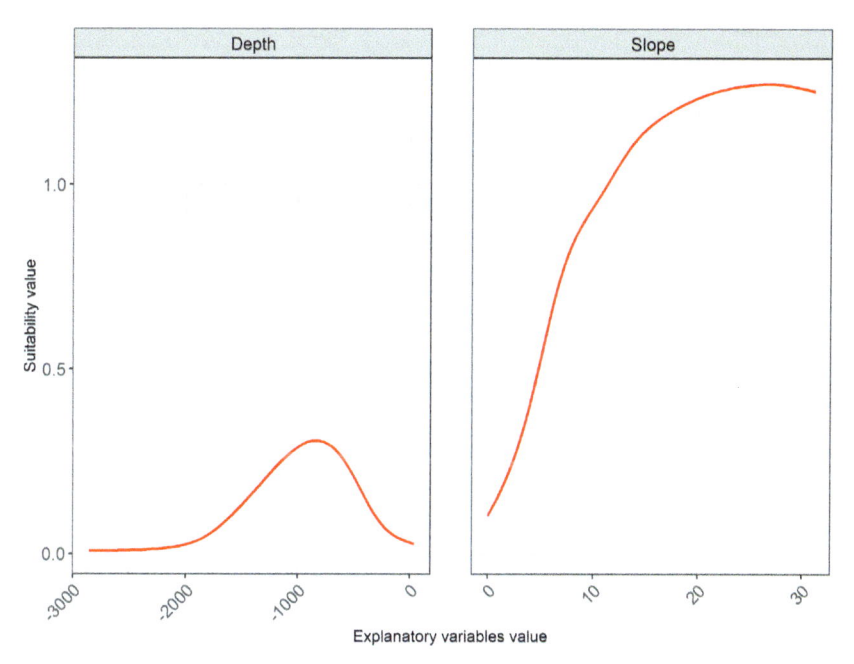

Fig. 5 Response curves of explanatory variables contributing to habitat suitability models for *K. sima* off the island of St. Vincent.

4. Discussion

This preliminary study reveals that the insular slope waters of the island of St. Vincent are an important habitat for dwarf sperm whales within the wider Caribbean region, where limited information exists on their occurrence, distribution, and habitat preferences (Cardona-Maldonado & Mignucci-Giannoni, 1999; Mutis & Polanco, 2019; De Weerdt et al., 2021). Similar studies conducted in peri-insular tropical and subtropical waters are relatively limited globally (e.g., Gannier, 2000; Anderson, 2005; Dunphy-Daly et al., 2008; Dulau-Drouot et al., 2008; Kiszka et al., 2010; Baird et al., 2013, 2021). Although survey design and conditions vary significantly between these regions, those studies suggest that encounter rates of dwarf sperm whales off St. Vincent are substantially higher than anywhere else yet identified. Indeed, encounter rates off St. Vincent are almost 20 times higher than off the big island of Hawai'i (Baird et al., 2021) and about four times higher than off Abaco in the Bahamas (MacLeod et al., 2004). Our results also suggest that the cetacean community of St. Vincent has a unique composition as 37.5% of all

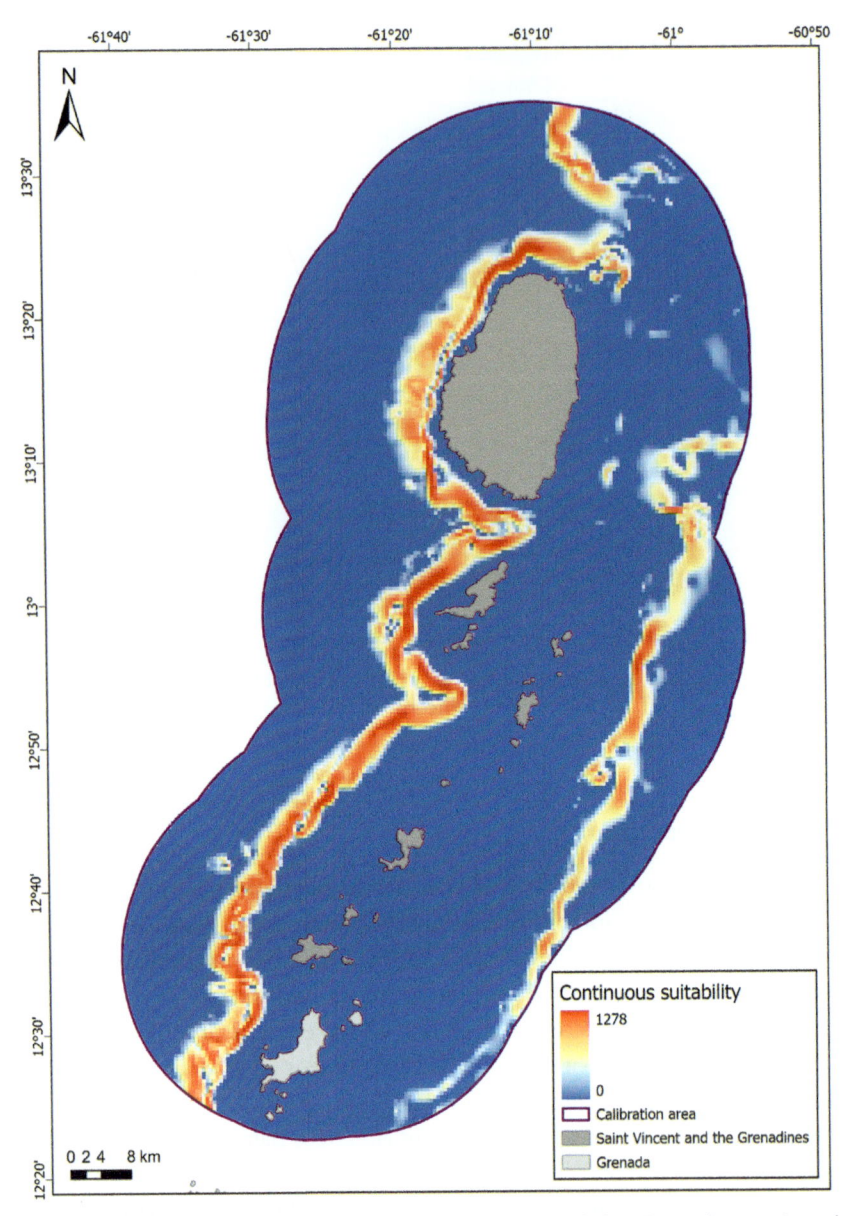

Fig. 6 Suitable habitats for *K. sima* off St. Vincent and the Grenadines using the ensemble model.

Fig. 7 Photographs of *K. sima* in the waters off the island of St. Vincent (May-June 2022 and 2023, credit: J. Kiszka, Florida International University). (A) adult female observed in June 2023 and exhibiting relative emaciation and multiple subcutaneous masses; (B) adult female with her calf (same individual as top left picture); (C) two adult dwarf sperm whales surfacing together (the individual on the foreground exhibits dark skin lesions, halfway between the blowhole and the dorsal fin, and the individual in the background exhibiting a linear scar on the dorsal fin, possible due to fishing line entanglement), (D) an adult dwarf sperm whale identified in May 2022.

cetacean sightings were dwarf sperm whales, whereas this proportion is 3.5% off the island of Hawai'i (Baird et al., 2021), for example. In addition, one of the most noticeable results that was obtained off St. Vincent is that encounter rates were not significantly influenced by sea state, unlike in other study regions (Dunphy-Daly et al., 2008; Baird et al., 2021). This is mostly likely due to the fact that at sea state conditions less than Beaufort 2 and 3, almost 90% of groups or individuals displayed aerial behaviors (full body breach) that enabled us to detect animals. These animals also breached relatively close to the research vessel (sometimes less than 50 m from the research vessel). We speculate that the impact of the bow of our research vessel on the swell or waves during surveys under these conditions generated a sound that alerted dwarf sperm whales, prompting them to jump out of the water to identify its source. It is also possible that beaching could be a behavior resulting from the irritation of the animals in the presence of the research vessel. However, high densities of these animals within a relatively small area might explain why sighting rates were high, despite the unfavorable sea conditions that we experienced during our two initial surveys, particularly as dwarf sperm whales are usually elusive (e.g., Baird et al., 2021).

Our visual survey data, collected off the west and south coast of St. Vincent, suggests that sightings are not uniformly distributed throughout the sampled area. All sightings occurred in the southwest, despite effort further north along the leeward side of the island (Fig. 1). This suggests that the core habitat of this species could be relatively narrow. Further effort around St. Vincent, including off the windward side of the island, particularly later in the summer when sea state is frequently below Beaufort 2, will be required to fully understand the fine-scale distribution of dwarf sperm whales. As seasons are known to have a significant influence on dwarf sperm whale habitat use and group size in other regions of the wider Caribbean, such as off Abaco in the Bahamas (Dunphy-Daly et al., 2008), future surveys will have to be stratified across the two main seasons (summer and winter, i.e. wet and dry seasons, respectively) off St. Vincent to investigate potential effects of seasons on their distribution and group dynamics.

Our results also suggest that dwarf sperm whales have a significant affinity for the upper portion of the insular slope, which is consistent with previous studies documenting their habitat preferences in the Bahamas (Dunphy-Daly et al., 2008), in the Gulf of Mexico (Ramírez-León et al., 2021), around the Mozambique Channel island of Mayotte (SW Indian Ocean; Kiszka et al., 2010), and off Hawai'i (Baird et al., 2021). Slope was found to be the most important predictor of habitat suitability of dwarf sperm whales off St. Vincent, which has been also reported in many other studies for a range of deep-diving whales, particularly sperm (*Physeter macrocephalus*) and beaked whales (Ziphiidae; Rogan et al., 2017; Virgili et al., 2021; Fiedler et al., 2023). Steep slope habitats along continental margins and around oceanic islands may promote prey aggregations (Logerwell & Smith, 2001; Vecchione, 2001), including pelagic cephalopods that are targeted by dwarf sperm whales throughout their range (Plön, 2004; Spitz et al., 2011; Staudinger et al., 2014; McAlpine, 2018).

Our data also provide some information on the behavior of dwarf sperm whales off St. Vincent. Group size and composition were similar to those reported at other locations, such as the Bahamas (e.g., Dunphy-Daly et al., 2008) or Hawai'i (Baird et al., 2021). However, we also report the largest group documented in the literature, with at least 20 traveling individuals observed in June 2023. The presence of such a large group further supports the assertion that the slope waters of St. Vincent represent an important habitat for this species at the regional and global level, and the drivers of grouping tactics will also have to be investigated. The limited photo-identification data collected off St. Vincent allowed to identify at least 22 distinct individuals, but no resighting was recorded (both within seasons and between years), which suggests that we identified a small proportion of the entire population, and/or that its range is much larger. Nevertheless, it confirms that photo-identification is a viable method to investigate the abundance, site fidelity, social structure, and other demographic parameters of dwarf sperm whales, as demonstrated by Baird et al. (2021). Photo-identification was also useful to document skin conditions in free-ranging dwarf sperm whales, which were not described in any other location. The origin of skin conditions observed on some individuals using photographs could not be determined. One of the individuals exhibited dark circular lesions that could possibly be associated with cell damage due to a poxvirus, which usually causes "tattoo skin disease" (Fig. 7C, individual in the foreground). However, histopathological analyses would be required to confirm this diagnostic. Unlike other locations, such as Hawai'i, no lesions associated with predator attempts by either killer whales (*Orcinus orca*) or large sharks were identified (Baird et al., 2021). However, local whalers targeting small cetaceans off St. Vincent regularly report that dwarf sperm whales are one of the most common prey for killer whales off St. Vincent (S. Hazelwood, personal communication, 3 February 2024). This is consistent with published information on the diet of killer whales in the Caribbean region, primarily consuming ocontocete cetaceans (Kiszka et al., 2021b). Future photo-identification efforts will allow for further evaluation of predator-prey interactions involving dwarf sperm whales, and investigation of the prevalence of skin diseases and therefore the health status of this population. Threats to dwarf sperm whales in this region include an active whaling operation and vessel traffic. Dwarf sperm whales are rarely targeted by artisanal whaling operations taking place off St. Vincent, primarily targeting short-finned pilot whales (*Globicephala macrorhynchus*) and other delphinids (Caldwell et al., 1973; Fielding & Kiszka, 2021). However, the maritime traffic is relatively important between the main island of SVG, particularly Bequia, which highly overlaps with the habitat where dwarf sperm whales occur. Ferries transit 12–16 times per day during daylight hours and, one of the sightings of dwarf sperm whales breaching in this study was in front of a ferry. Future research will be needed to investigate the effects of noise pollution on this population, particularly during the day when these animals mostly rest at the surface.

Acknowledgments

Funding and support for this project was provided by the National Science Foundation (Grant No. #1827195, awarded to M. Heithaus, J. Kiszka and C. Gomes) and from the FIU Institute of Environment's Coastlines and Oceans Division travel grant. We are also grateful to the Fisheries Division of the Ministry of Agriculture, Forestry, Fisheries, Rural Transformation, Industry and Labor of St. Vincent and the Grenadines for their support. We also thank students involved in the data collection, particularly Sophia Hemsi, Amanda Di Perna, Colleen Rodriguez, Alexandra Pfeiffer, and our boat pilots, particularly Hal Daize and Rodney Dabreo. The authors also thank Dr. Ruth Ewing (NOAA Southeast Fisheries Science Center) and Dr. Rocio Gonzalez Barrientos (Texas A&M University) for providing their expertise to identify the skin lesions observed on dwarf sperm whales. The authors are also grateful to Stephanie Plön, Robin W. Baird, and two anonymous reviewers for their constructive comments on the first version of the manuscript. This is contribution #1742 from the Institute of Environment at Florida International University. This manuscript is dedicated to the memory of Hal Daize, a prioneer of whale watching activities in St. Vincent.

Appendix A. Supporting information

Supplementary data associated with this article can be found in the online version at https://doi.org/10.1016/bs.amb.2024.09.002.

References

Aiello-Lammens, M.E., Boria, R.A., Radosavljevic, A., Vilela, B., Anderson, R.P., 2015. spThin: an R package for spatial thinning of species occurrence records for use in ecological niche models. Ecography 38, 541–545.

Anderson, R.C., 2005. Observations of cetaceans in the Maldives, 1990-2002. J. Cetacean Res. Manag. 7, 119–135.

Baird, R.W., 2005. Sightings of dwarf (*Kogia sima*) and pygmy (*K. breviceps*) sperm whales from the main Hawaiian Islands. Pac. Sci. 59, 461–466.

Baird, R.W., Mahaffy, S.D., Lerma, J.K., 2021. Site fidelity, spatial use, and behavior of dwarf sperm whales in Hawaiian waters: using small-boat surveys, photo-identification, and unmanned aerial systems to study a difficult-to-study species. Mar. Mammal. Sci. 38, 326–348.

Baird, R.W., Webster, D.L., Aschettino, J.M., Schorr, G.S., McSweeney, D.J., 2013. Odontocete cetaceans around the main Hawaiian Islands: habitat use and relative abundance from small-boat sighting surveys. Aquat. Mamm. 39, 253–260.

Barlow, J., 2015. Inferring trackline detection probabilities, g (0), for cetaceans from apparent densities in different survey conditions. Mar. Mammal. Sci. 31, 923–943.

Caldwell, D.K., Caldwell, M.C., Arrindell, G., 1973. Dwarf sperm whales, *Kogia simus*, from the Lesser Antillean island of St. Vincent. J. Mammal. 54, 515–517.

Cañadas, A., Sagarminaga, R., Garcıa-Tiscar, S., 2002. Cetacean distribution related with depth and slope in the Mediterranean waters off southern Spain. Deep. Sea Res. Part. I: Oceanographic Res. Pap. 49, 2053–2073.

Cardona-Maldonado, M.A., Mignucci-Giannoni, A.A., 1999. Pygmy and dwarf sperm whales in Puerto Rico and the Virgin Islands, with a review of *Kogia* in the Caribbean. Caribb. J. Sci. 35, 29–37.

Chivers, S.J., Leduc, R.G., Robertson, K.M., Barros, N.B., Dizon, A.E., 2005. Genetic variation of *Kogia* spp. with preliminary evidence for two species of *Kogia sima*. Mar. Mammal. Sci. 21, 619–634.

De Weerdt, J., Ramos, E.A., Pouplard, E., Kochzius, M., Clapham, P., 2021. Cetacean strandings along the Pacific and Caribbean coasts of Nicaragua from 2014 to 2021. Mar. Biodivers. Rec. 14, 1–9.

Logerwell, E.A., Smith, P.E., 2001. Mesoscale eddies and survival of late stage Pacific sardine (*Sardinops sagax*) larvae. Fish. Oceanogr. 10, 13–25.
MacLeod, C.D., Hauser, N., Peckham, H., 2004. Diversity, relative density and structure of the cetacean community in summer months east of Great Abaco, Bahamas. J. Mar. Biol. Assoc. U Kingd. 84, 469–474.
McAlpine, D.F., 2018. Pygmy and dwarf sperm whales *Kogia breviceps* and *K. simus*. In: Wursig, B., Thewissen, J.G.M., Kovacs, K.M. (Eds.), Encyclopedia of Marine Mammals, third ed.,. Academic Press., pp. 786–788.
Moura, J.F., Acevedo-Trejos, E., Tavares, D.C., Meirelles, A.C., Silva, C.P., Oliveira, L.R., et al., 2016. Stranding events of *Kogia* whales along the Brazilian coast. PLoS One 11, e0146108.
Mutis, M.A., Polanco, A., 2019. First stranding record of *Kogia sima* (Owen, 1866) in the Colombian Caribbean. Lat. Am. J. Aquat. Mamm. 14, 18–26.
Odell, D.K., 1991. A review of the Southeastern United States Marine Mammal Stranding Network: 1978–1987. Marine mammal strandings in the United States, pp. 19–23. NOAA Technical Report NMFS 98.
Palka, D.L., 2012. Cetacean abundance estimates in US northwestern Atlantic Ocean waters from summer 2011 line transect survey. Northeast Fisheries Science Center Reference Document. NOAA, pp. 12–29.
Phillips, S.J., Anderson, R.P., Schapire, R.E., 2006. Maximum entropy modeling of species geographic distributions. Ecol. Model. 190, 231–259.
Phillips, S.J., Dudík, M., 2008. Modeling of species distributions with Maxent: new extensions and a comprehensive evaluation. Ecography 31, 161–175.
Plön, 2004. The Status and Natural History of Pygmy (*Kogia breviceps*) and Dwarf (*K. sima*) Sperm Whales off Southern Africa. PhD thesis. Rhodes. University, Grahamstown, South Africa, pp. 553.
Plön, 2022. Pygmy sperm whale *Kogia breviceps* (de Blainville, 1838). In: Hackländer, K., Zachos, F.E. (Eds.), Handbook of the Mammals of Europe. Springer, Cham. https://doi.org/10.1007/978-3-319-65038-8_90-1.
Plön, S., Baird, R.W., 2022. Dwarf sperm whale, *Kogia sima* (Owen, 1866). Handbook of the Mammals of Europe. Springer International Publishing,, Cham, pp. 1–14.
Plön, S., Best, P.B., Duignan, P., Lavery, S.D., Bernard, R.T.F., Van Waerebeek, K., et al., 2023. Population structure of pygmy (*Kogia breviceps*) and dwarf (*Kogia sima*) sperm whales in the Southern Hemisphere may reflect foraging ecology and dispersal patterns. Adv. Mar. Biol. 96, 85–114.
Ramírez-León, M.R., García-Aguilar, M.C., Romo-Curiel, A.E., Ramírez-Mendoza, Z., Fajardo-Yamamoto, A., Sosa-Nishizaki, O., 2021. Habitat suitability of cetaceans in the Gulf of Mexico using an ecological niche modeling approach. PeerJ 9, e10834.
Redfern, J.V., Ferguson, M.C., Becker, E.A., Hyrenbach, K.D., Good, C., Barlow, J., et al., 2006. Techniques for cetacean–habitat modeling. Mar. Ecol. Prog. Ser. 310, 271–295.
Rogan, E., Cañadas, A., Macleod, K., Santos, M.B., Mikkelsen, B., Uriarte, A., et al., 2017. Distribution, abundance and habitat use of deep diving cetaceans in the North-East Atlantic. Deep. Sea Res. Part. II: Topical Stud. Oceanogr. 141, 8–19.
Spitz, J., Cherel, Y., Bertin, S., Kiszka, J., Dewez, A., Ridoux, V., 2011. Prey preferences among the community of deep-diving odontocetes from the Bay of Biscay, Northeast Atlantic. Deep. Sea Res. Part. I: Oceanogr. Res. Pap. 58, 273–282.
Staudinger, M.D., McAlarney, R.J., McLellan, W.A., Pabst, D.A., 2014. Foraging ecology and niche overlap in pygmy (*Kogia breviceps*) and dwarf (*Kogia sima*) sperm whales from waters of the US mid-Atlantic coast. Mar. Mammal. Sci. 30, 626–655.
Syme, J., Kiszka, J.J., Parra, G.J., 2022. How to define a dolphin "group"? Need for consistency and justification based on objective criteria. Ecol. Evolution 1 12, e9513.

Dulau-Drouot, V., Boucaud, V., Rota, B., 2008. Cetacean diversity off La Réunion Island (France). J. Mar. Biol. Assoc. U Kingd. 88, 1263–1272.

Dunphy-Daly, M.M., Heithaus, M.R., Claridge, D.E., 2008. Temporal variation in dwarf sperm whale (*Kogia sima*) habitat use and group size off Great Abaco Island, Bahamas. Mar. Mammal. Sci. 24, 171–182.

Fiedler, P.C., Becker, E.A., Forney, K.A., Barlow, J., Moore, J.E., 2023. Species distribution modeling of deep-diving cetaceans. Mar. Mammal. Sci. 39, 1178–1203.

Fielding, R., Kiszka, J.J., 2021. Artisanal and aboriginal subsistence whaling in Saint Vincent and the Grenadines (Eastern Caribbean): history, catch characteristics, and needs for research and management. Front. Mar. Sci. 8, 668597.

Fire, S.E., Wang, Z., Leighfield, T.A., Morton, S.L., McFee, W.E., McLellan, W.A., et al., 2009. Domoic acid exposure in pygmy and dwarf sperm whales (*Kogia* spp.) from southeastern and mid-Atlantic US waters. Harmful Algae 8, 658–664.

Gannier, A., 2000. Distribution of cetaceans off the Society Islands (French Polynesia) as obtained from dedicated surveys. Aquat. Mamm. 26, 111–126.

Garrison, L.P., 2016. Abundance of marine mammals in waters of the US east coast during the summer 2011, Southeast Fisheries Science Center Reference Document PRDB-2016-08.

Garrison, L.P., Martinez, A., Maze-Foley, K., 2010. Habitat and abundance of cetaceans in Atlantic Ocean continental slope waters off the eastern USA. J. Cetacean Res. Manag. 11, 67–277.

Guisan, A., Thuiller, W., Zimmermann, N.E., 2017. Habitat Suitability and Distribution Models: with Applications in R. Cambridge University Press.

Hodge, L.E., Baumann-Pickering, S., Hildebrand, J.A., Bell, J.T., Cummings, E.W., Foley, H.J., et al., 2018. Heard but not seen: occurrence of *Kogia* spp. along the western North Atlantic shelf break. Mar. Mammal. Sci. 34, 1141–1153.

Ilangakoon, A.D., 2012. A review of cetacean research and conservation in Sri Lanka. J. Cetacean Res. Manag. 12, 177–183.

Kiszka, J., Braulik, G., 2020. *Kogia sima*, dwarf sperm whale. The IUCN Red List of Threatened Species. e.T11048A50359330. https://dx.doi.org/10.2305/IUCN.UK.2020-2.RLTS.T11048A50359330.en.

Kiszka, J., Braulik, G., 2022. *Kogia breviceps*, pygmy sperm whale. The IUCN Red List of Threatened Species 2020: e.T11047A50358334. https://dx.doi.org/10.2305/IUCN.UK.2020-2.RLTS.T11047A50358334.en.

Kiszka, J.J., Caputo, M., Méndez-Fernandez, P., Fielding, R., 2021b. Feeding ecology of elusive Caribbean killer whales inferred from Bayesian stable isotope mixing models and whalers' ecological knowledge. Front. Mar. Sci. 8, 648421.

Kiszka, J., Ersts, P.J., Ridoux, V., 2010. Structure of a toothed cetacean community around a tropical island (Mayotte, Mozambique Channel). Afr. J. Mar. Sci. 32, 543–551.

Kiszka, J.J., Moazzam, M., Boussarie, G., Shahid, U., Khan, B., Nawaz, R., 2021a. Setting the net lower: a potential low-cost mitigation method to reduce cetacean bycatch in drift gillnet fisheries. Aquat. Conservation: Mar. Freshw. Ecosyst. 31, 3111–3119.

Kiszka, J., Pelourdeau, D., Ridoux, V., 2008. Body scars and dorsal fin disfigurements as indicators interaction between small cetaceans and fisheries around the Mozambique Channel island of Mayotte. West. Indian. Ocean. J. Mar. Sci. 7, 185–193.

Laran, S., Authier, M., Van Canneyt, O., Dorémus, G., Watremez, P., Ridoux, V., 2017. A comprehensive survey of pelagic megafauna: their distribution, densities, and taxonomic richness in the tropical Southwest Indian Ocean. Front. Mar. Sci. 4, 139.

Lettrich, M.D., Asaro, M.J., Borggaard, D.L., Dick, D.M., Griffis, R.B., Litz, J.A., et al., 2023. Vulnerability to climate change of United States marine mammal stocks in the western North Atlantic, Gulf of Mexico, and Caribbean. PLoS One 18, e0290643.

Thuiller, W., Georges, D., Gueguen, M., Engler, R., Breiner, F., Lafourcade, B., et al., 2023. biomod2: ensemble platform for species distribution modeling. R. Package Version 4, 2–5.

Van Bressem, M.F.E., Flach, L., Reyes, J.C., Echegaray, M., Santos, M., Viddi, F., et al., 2015. Epidemiological characteristics of skin disorders in cetaceans from South American waters. Lat. Am. J. Aquat. Mamm. 10, 20–32.

Van Bressem, M.F., Van Waerebeek, K., Aznar, F.J., Raga, J.A., Jepson, P.D., Duignan, P., et al., 2009. Epidemiological pattern of tattoo skin disease: a potential general health indicator for cetaceans. Dis. Aquat. Org. 85, 225–237.

van Weelden, C., Towers, J.R., Bosker, T., 2021. Impacts of climate change on cetacean distribution, habitat and migration. Clim. Change Ecol. 1, 100009.

Vecchione, M., 2001. Cephalopods of the Continental Slope. Am. Fish. Soc. Symposium 25, 153–160.

Virgili, A., Hedon, L., Authier, M., Calmettes, B., Claridge, D., Cole, T., et al., 2021. Towards a better characterisation of deep-diving whales' distributions by using prey distribution model outputs? PLoS One 16, e0255667.

Yong, W.C., Chou, L.S., Jepson, P.D., Brownell Jr, R.L., Cowan, D., Chang, P.H., et al., 2008. Unusual cetacean mortality event in Taiwan, possibly linked to naval activity. Vet. Rec. 162, 184–186.

Zurell, D., Franklin, J., König, C., Bouchet, P.J., Dormann, C.F., Elith, J., et al., 2020. A standard protocol for reporting species distribution models. Ecography 43, 1261–1277.

CHAPTER TWO

Strandings and at sea observations reveal the canary archipelago as an important habitat for pygmy and dwarf sperm whale

Vidal Martín[a,*], Marisa Tejedor[b], Manuel Carrillo[c,1], Mónica Pérez-Gil[d], Manuel Arbelo[e], Antonella Servidio[d], Enrique Pérez-Gil[d], Nuria Varo-Cruz[d], Francesca Fusar Poli[a], Sol Aliart[a], Gustavo Tejera[f], Marta Lorente[a], and Antonio Fernández[e]

[a]Society for the Study of Cetacean in the Canary Archipelago (SECAC), Canary Islands Cetacean Research Centre, Canary Islands Stranding Network, Lanzarote, Canary Islands, Spain
[b]Canary Islands Cetaceans Stranding Network, Playa Blanca, Lanzarote, Canary Islands, Spain
[c]Canarias Conservación, La Laguna, Tenerife, Spain
[d]Cetacean and Marine Research Institute of the Canary Islands (CEAMAR), Playa Honda, Lanzarote, Canary Islands, Spain
[e]Veterinary Histology and Pathology, Atlantic Center for Cetacean Research, University Institute of Animal Health and Food Safety (IUSA), Veterinary School, University of Las Palmas de Gran Canaria, Canary Islands, Spain
[f]Canary Islands´ Ornithology and Natural History Group (GOHNIC)
*Corresponding author. e-mail address: vidal@cetaceos.org

Contents

1. Introduction	22
2. Material and methods	25
2.1 Study area	25
2.2 Strandings	25
2.3 Sightings at sea	26
2.4 Species identification	27
3. Results	28
3.1 Strandings	28
3.2 Sightings	38
4. Discussion	47
4.1 Strandings	47
4.2 Reproduction	49
4.3 Presence in the Canary Archipelago	50
4.4 Predation	52
4.5 Conservation	54

[1] Deceased author.

5. Conclusions	56
6. Future perspectives	57
Acknowledgments	58
References	58

Abstract

Cetaceans are a critical component of marine ecosystems, acting as top predators in mesopelagic trophic webs. In the Macaronesian biogeographical region, cetacean populations face threats from various anthropogenic activities. Evaluating cryptic oceanic species like kogiids whales is challenging due to insufficient biological and ecological data, making conservation assessments and management efforts harder to achieve. *Kogia breviceps* and *K. sima* comprising the family Kogiidae, are morphologically similar, widely distributed, and elusive, with most information originating from stranded specimens and few at sea observations. This study examines data from *Kogia* species stranded in the Canary Islands between 1977 and 2024 and analyzes sighting data obtained between 1999 and 2024. Between 1977 and May 2024, there were 111 stranding events involving 114 kogiid individuals along the Canary Islands' coasts: 86 events (88 individuals) were pygmy sperm whales, 14 events (15 individuals) were dwarf sperm whales, and 11 events with 11 individuals, were unidentified *Kogia* species. Additionally, 36 kogiid sightings were recorded, of which 34 originated from dedicated surveys and 2 from opportunistic sightings. Of these sightings, 14 (39%) were *K. breviceps*, 9 (25%) were *K. sima*, and 13 (36%) were unidentified *Kogia*. Twenty-nine sightings (80.5%) of kogiids were recorded in the waters off the eastern coast of the islands of Lanzarote and Fuerteventura. The data indicate that the waters around the Canary Islands are an important habitat for *Kogia* whales. The findings establish a baseline for future research and underscore the necessity of accurately assessing conservation pressures on pygmy and dwarf sperm whales in the region.

1. Introduction

Cetaceans are part of the marine megafauna, being an essential ecological component in oceanic ecosystems as apex and meso-predators in pelagic and coastal food webs, prey of other predators, carbon reservoirs, vertical and horizontal vector of nutrients, and detrital sources for the deep-sea (Roman et al., 2014; Albouy et al., 2020; Norris et al., 2020). In addition, they are indicators of ocean health. The Canary Archipelago constitutes a hotspot of cetacean diversity, encompassing 31 species belonging to seven families, 32.9% of the cetacean species recorded around the world (List of Marine Mammals Species and Subspecies, Committee of Taxonomy of the Society for Marine Mammalogy, consulted on 18 June 2024). Due to its geographic position, climatic conditions, and highly dynamic oceanographic processes, the archipelago promotes enhanced productivity and trophic resources in pelagic habitats, both at the surface and in deep waters (Doty and Oguri, 1956; Mann and Lazier, 1991). Oceanic island systems represent essential habitats for deep-diving cetaceans, most of which are poorly known taxa due to their elusive, cryptic, and remote nature. In the Canary archipelago, several Sites of Community Interest (SCIs) and Special Areas of Conservation (SACs) have been designated as part of the Natura 2000 Network, implementing Directive 92/43/EEC on the Conservation of Natural Habitats and of Wild Fauna and Flora (Council Directive 92/43/EEC of 21 May 1992), and the Marine Strategy Framework Directive (Directive 2008/56/EC). Connectivity in the movements of several cetacean species has been evidenced between the insular systems in the Macaronesia biogeographical region, especially between neighboring Madeira and the Canary Islands (Alves et al., 2019; Dinis et al., 2021; Ferreira et al., 2022; Gómez-Lobo et al., 2024).

Cetacean populations in the insular systems of the Macaronesia region are vulnerable to various anthropogenic activities (McLuor et al., 2022). A common issue in assessing conservation risk of oceanic cetaceans is the lack of adequate data on fundamental aspects of their natural history and their distribution range, hindering objective assessments of the conservation status. In this context, stranded cetaceans constitute a valuable source of scientific information for these species relevant to their conservation and population management (Coombs et al., 2019). Strandings are especially important for rare and inconspicuous species that are poorly known, such as pygmy and dwarf sperm whales or beaked whales (Thompson et al., 2013). The Canary Islands have had an alert and response stranding network (Canary Islands Stranding Network) for more than 30 years, coordinated by the Government of the Canary Islands, that includes a scientific program focused on the health and biological aspects of stranded animals.

The family Kogiidae is represented by two sister species: *Kogia breviceps* (pygmy sperm whale) and *Kogia sima* (dwarf sperm whale), forming part of the superfamily Physeteroidea and subfamily Kogiinae (Heyning, 1997; Rice, 1998). Kogiids are deep-diving cetaceans, performing prolonged dives of up to 43 min (Breese and Tershey, 1993). They are widely distributed in temperate to tropical oceanic waters of all oceans, with a sympatric distribution over much of their range. Although, the dwarf sperm whale prefers warmer waters than the pygmy sperm whale (Caldwell and Caldwell, 1989; Willis and Baird, 1998; Plön 2022; Plön et al., 2023).

Although kogiids frequently strand in certain regions, this group is among the rarest odontocetes seen due to the difficulty of studying them at sea. Pygmy and dwarf sperm whales are difficult to detect at sea due to their small size, usually indistinguishable blow, low-profile at the surface, inconspicuous and elusive behavior, and brief sighting times (Scott and Cordaro, 1987; Baird et al., 2021; Plön 2022; Plön and Baird, 2022). They are found alone or in small groups (Nagorsen, 1985; McAlpine, 2018; Baird et al., 2021; Plön 2022; Plön and Baird, 2022). The probability of sighting depends on sea conditions and increases with Beaufort 0 (Barlow, 2015). The risk of confusion in identifying the two species at sea is high because they share similar morphological characteristics, especially in calves and immature individuals (McAlpine, 2014, 2018; Jefferson et al., 2015; Baird et al., 2021). In addition, pygmy and dwarf sperm whales cannot be acoustically distinguished at the species level (Hildebrand et al., 2019). Thus, most of what is known about their natural history is inferred from strandings (Ogawa, 1936; Yamada, 1954; Handley, 1966; Best, 2007; McAlpine, 2018; Baird et al., 2021; Plön 2022; Plön and Baird, 2022). They can be differentiated based on maximum body length, the position and height of the dorsal fin, head morphology (McAlpine, 2018), gill coloration (Keenan-Bateman et al., 2016), and several skull and postcranial characters (Ross, 1979, 1984; Chivers et al., 2005). Although the position and height of the dorsal fin are diagnostic features for discriminating between the two species, this morphological characteristic is variable (Plön, 2004; Best, 2007; Yamada, 1954; Baird et al., 2021). Limited studies at sea have been realized in a few locations worldwide where dwarf sperm whales maintain stable populations, such as Hawai'i (Baird et al., 2021) and the Bahamas (Dunphy-Daly et al., 2008).

Both species have been recorded in the Canary Islands. The first record of a pygmy sperm whale (Casinos, 1977) is of a stranded live animal on a beach near an undetermined location north of Gran Canaria on February 12, 1973. Mr. John Nuncio from Durban, South Africa, found and photographed this specimen and sent the photographs to Dr. P. Best. The species confirmation was conducted by Dr. P. Best, P.J.H. Van Bree, R. Duguy, and C. F. Fraser. The first record of a dwarf sperm whale was through a partial skull and skeleton collected by Dr. R. Hutterer from the beach of Cofete (UTM grid ES5709), Fuerteventura, on March 6, 1987 (Hutterer, 1994). The skull of this specimen is in the mammal collection of the Museum Koenig in Bonn, Germany (ZFMK 87.723). Maritime traffic and vessel collisions with both species have been reported in the Canary

Islands and other areas and Kogiids have been involved in stranding events associated with navy mid-frequency active antisubmarine sonar use in the Canary Islands and elsewhere (Simmonds and Lopez-Jurado, 1991; Hohn et al., 2006; Arregui et al., 2019). This study provides an initial comprehensive assessment of the existing knowledge on pygmy and dwarf sperm whales in the Canary Islands based on at-sea sightings and stranding records to the present date.

2. Material and methods
2.1 Study area

The data for the present study were collected in the Canary Islands (from 29° 24′ 40″ N to 27° 58′ 16″ N and from 13° 19′ 54″ W to 18° 09′ 38″ W), an archipelago of volcanic origin located in the subtropical northeast Atlantic Ocean and situated between 100 and 600 km off the western African continental edge, consisting of eight main islands: La Graciosa, Lanzarote, Fuerteventura, Gran Canaria, Tenerife, La Gomera, La Palma, and El Hierro. Due to their volcanic origin, the underwater morphology of the Canary Islands is sharply defined, with very narrow insular shelves and steep slopes (particularly in the westernmost islands), reaching depths of up to 3000 m in the channels between the main islands and 1300 m between the eastern islands and the neighboring African coast. Oceanographic conditions in the region are influenced by the Cold Canary Current, which represents the descending branch of the Northeast Subtropical Atlantic Gyre, the upwelling waters of the northwestern African coast, and several mesoscale oceanographic phenomena (Barton et al., 1998; Aristegui et al., 1994; Ariza et al., 2016). The Canary Basin is part of the Canary Current Large Marine Ecosystem (CCLME), characterized by high marine productivity and rich biodiversity (Vélez-Belchí et al., 2015).

2.2 Strandings

The study of stranded cetaceans in the Canary Islands began in the mid-1980s, with an early stranding network established in the early 1990s. The current official framework of the Canary Islands Stranding Network was constituted in 2000, with logistical coordination by the Canary Islands Government. Minimal information collected from stranded cetaceans included location, date, standard length (SL), sex, carcass classification or degree of decomposition, maturity (sexual, cranial, physical), and standard external morphometrics following Norris (1961). Colouration patterns were evaluated through direct observations or photographs taken of fresh specimens. The conservation condition of the carcasses was classified as very fresh, fresh, moderate autolysis, advanced autolysis, or very advanced autolysis (Kuiken and García-Hartmann, 1991; Arbelo et al., 2013; Díaz-Delgado et al., 2018). According to Joblon et al. (2014), body condition was subjectively classified on the external physical appearance into four categories: good, moderate, poor, and emaciated. Age categories were established based on total body length and histological gonadal examination according to Geraci and Lounsbury (2005) as neonates or newborns (animals with vibrissae or vibrissae crypts, unhealed umbilicus, fetal folds, soft and folded over dorsal fin and tail flukes), calves (presence of milk in the stomach or about the size of a nursing calf), juveniles (sexually immature animal with a body length smaller than an adult, but larger than a calf), subadults (body length of an adult, but immature gonads), and adults (mature gonads). Postmortem studies were conducted on the carcasses, ranging from basic biological dissections to standardized forensic necropsies conducted by a team of the University Institute of Animal Health (IUSA) of the University of Las Palmas de Gran Canaria (ULPGC). Biological information recorded included the collection of stomach contents, reproductive organs, epibionts, parasites, and tissue samples for genetic and trophic studies. Sexual maturity was determined based on histological examination of the testes in males, the presence of ovarian scars in females, as well as lactation, pregnancy, and the presence of a dependent calf. Whenever possible, the specimens' osteological material (skull and postcranial skeleton) was collected and deposited in the Biological Reference Collection for Cetaceans of Macaronesia (BRCCM) overseen by the Society for the Study of Cetaceans in the Canary Archipelago (SECAC) or other scientific institutions in the Canary Islands.

2.3 Sightings at sea

Sightings of *Kogia* species were recorded during dedicated surveys conducted between 1999 and 2024 as part of a multi-species cetacean study by SECAC and the Cetaceans and Marine Research Institute for the Canary Islands (CEAMAR) in the Canary Archipelago, along with a few opportunistic observations from citizen science, verified through documentary evidence, such as photographs and videos. Dedicated surveys were conducted in different periods and islands using various research vessels, including small semi-rigid inflatable boats, motor vessels, sailing boats, and occasionally opportunistic platforms such as whale-watching boats and ferries. The methodology was similar in all cases. The sampling effort followed predetermined and non-predetermined track lines in a zig-zag pattern from the coastline to offshore, covering the entire bathymetric range and habitats of the study area. The boats traveled at a speed of 8.1–10.8 knots, covering approximately up to 12 nautical miles offshore from the coast of each island and the channels between islands. Surveys were conducted when the sea state was less than three on the Beaufort scale. During each survey, at least two trained observers scanned the horizon, covering a 180-degree sector in front of the boat from the highest available position, alternating between naked eye and 7 × 50 binoculars. Data on sightings, effort, and environmental data were collected, including sea state (Beaufort and Douglas scale), visibility, and a subjective condition index to detect cetaceans based on the combination of different environmental variables. Whenever a cetacean or a group of cetaceans was detected, the search effort was interrupted and the animal was approached for species identification, with sighting data being collected. The distance to the animals at the moment of detection was calculated using the reticle of the binoculars or estimated visually. Time, GPS position, estimated radial distance, initial cue, tentative species, behavior, and group size were recorded for each encounter. When possible, photographs were taken using digital SLR cameras to confirm species identification, which was initially done at sea and later verified through the photographs obtained during the encounters. Due to sea and wind conditions, the survey effort was concentrated in the southwestern sectors off the islands of Gran Canaria, Tenerife, La Gomera, La Palma, and El Hierro, as well as along the eastern coasts of Lanzarote and Fuerteventura, at the expense of the inter-island channels and the leeward-facing coasts of the islands, where less effort was expended.

2.4 Species identification

Species identity of the stranded animals was realized based on the morphological features of the carcasses and, when possible, through cranial characteristics proposed by various authors (Handley, 1966; Ross, 1979, 1984; Caldwell and Caldwell, 1989; Chivers et al., 2005; McAlpine, 2014, 2018). To ensure accurate species identification, authors with previous experience with identification of *Kogia* species (VM, MC, MT, MP, EP, NV, MA, AF) conducted a retrospective review of all stranding cases, examining available material from the animals, including those not studied *in situ* by us, but with existing evidence, such as photographs. To distinguish between both species at sea, photographs were examined for diagnostic features, such as dorsal fin position and height, relative size of the animals, and the curvature of the back (usually visible in pygmy sperm whales). The stranded animals for which there was insufficient information for diagnosis or sea observations without photographs were assigned as unidentified *Kogia*.

3. Results
3.1 Strandings

Between 1977 and 2024, there were 111 stranding events involving 114 individuals of kogiids in the Canary Islands. Of these, 86 events involving 88 individuals were pygmy sperm whales, 14 events (15 individuals) were dwarf sperm whales, and 11 events with 11 individuals were of unidentified *Kogia* species. Kogiids represented 7.27% of the 1526 cetacean stranding events recorded in the Canary Archipelago between 1977 and 2024. Fig. 1 presents the annual strandings of pygmy, dwarf, and unidentified *Kogia* spp. across different time intervals during the study period in the Canary Islands.

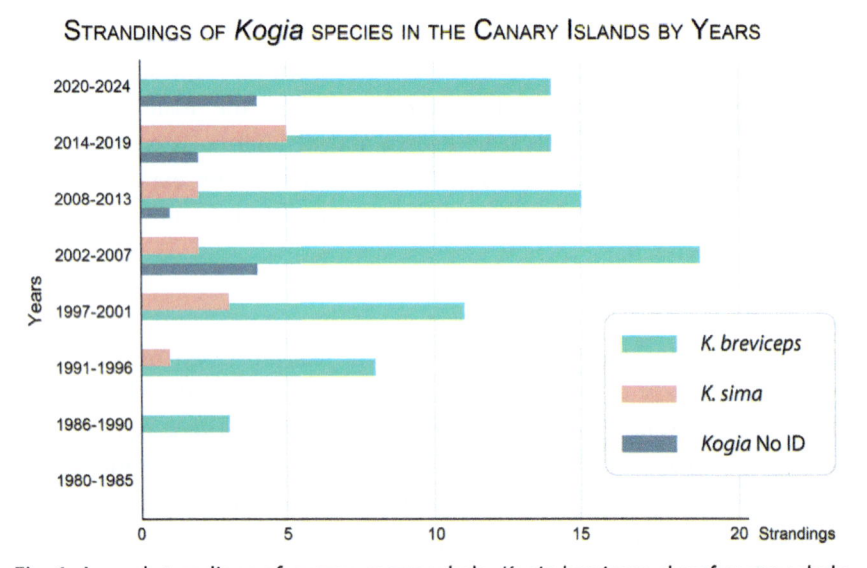

Fig. 1 Annual strandings of pygmy sperm whale, *Kogia breviceps*, dwarf sperm whale, *K. sima*, and unidentified kogiids during the study period in the Canary Islands.

3.1.1 Pygmy sperm whale

Stranding cases consisted of 20 events (23.8%) with 24 animals as live strandings (active), 62 events (73.8%) with 62 animals as dead strandings (passive), and two events (2.4%) with two animals as indeterminate. Five live-stranded animals were returned to the sea, including a 2.4 m female stranded on May 21, 2001, in Gran Canaria (Kbr23), which was kept in a tank for 11 days before being returned to the sea. Pygmy sperm whale strandings accounted for 5.63% of the total cetacean stranding events in the Canary Islands between 1977 and 2024. For pygmy sperm whales, six animals out of a total of 83 carcasses were not examined: subsequent post-mortem studies were conducted on 77 specimens. The spatial distribution of stranding events of pygmy sperm whales was: Lanzarote and La Graciosa ($n = 15$; 18.1%), Fuerteventura ($n = 24$; 28.9%), Gran Canaria ($n = 17$; 20.5%), Tenerife ($n = 22$; 26.5%), La Gomera ($n = 5$; 6.0%), and La Palma ($n = 1$; 1.2%; Fig. 2A). Strandings have been recorded throughout all months, with a monthly event ranging from 3 to 11 animals (mean ± SD = 7 ± 2.09), showing a slight peak in winter and spring (Fig. 3A). Carcass condition included very fresh and fresh ($n = 23$; 27.7%), moderate autolysis ($n = 8$; 9.6%), advanced autolysis ($n = 14$; 16.9%), and very advanced autolysis ($n = 36$; 43.4%), or was indeterminate ($n = 2$; 2.4%). The sex distribution of all individuals (including the refloated animals) was 29 (33.0%) females, 44 (50.0%) males, and 15 (17.0%) indeterminate. The age categories for all individuals (including the refloated animals) were 4 (4.5%) neonates, 14 (15.9%) calves, 7 (8.0%) juveniles, 46 (52.3%) adults, and 17 (19.3%) indeterminate.

The length of males ranged from 152 cm to 343 cm (mean ± SD = 263.3 ± 52.8; $n = 42$), and females from 110 cm to 302 cm (mean ± SD = 237.7 ± 53.9; $n = 27$) (Fig. 4). Only a small number of animals were weighed due to logistical limitations. Table 1 presents the minimum and maximum weights recorded for pygmy sperm whales of both sexes stranded in the Canary Archipelago during the study period. The length of mature males ranged from 246 cm to 343 cm (mean ± SD = 294.1 ± 28.9; $n = 26$), and the length of mature females ranged from 260 cm to 302 cm (mean ± SD = 2.81,85 ± 11.9; $n = 13$). The combined weight (with epididymides) of the testicles of the mature animals ranged from 1640 g to 7500 g, with a median of 3180 g (mean ± SD = 3649.71 ± 2138.71; $n = 7$). The right testicle length of the adult animals ranged from 370 to 630 mm (mean ± SD = 462.46 mm ± 69.89; $n = 13$), and the left testicle length ranged from 340 to 590 mm, with a median of 452.62 mm

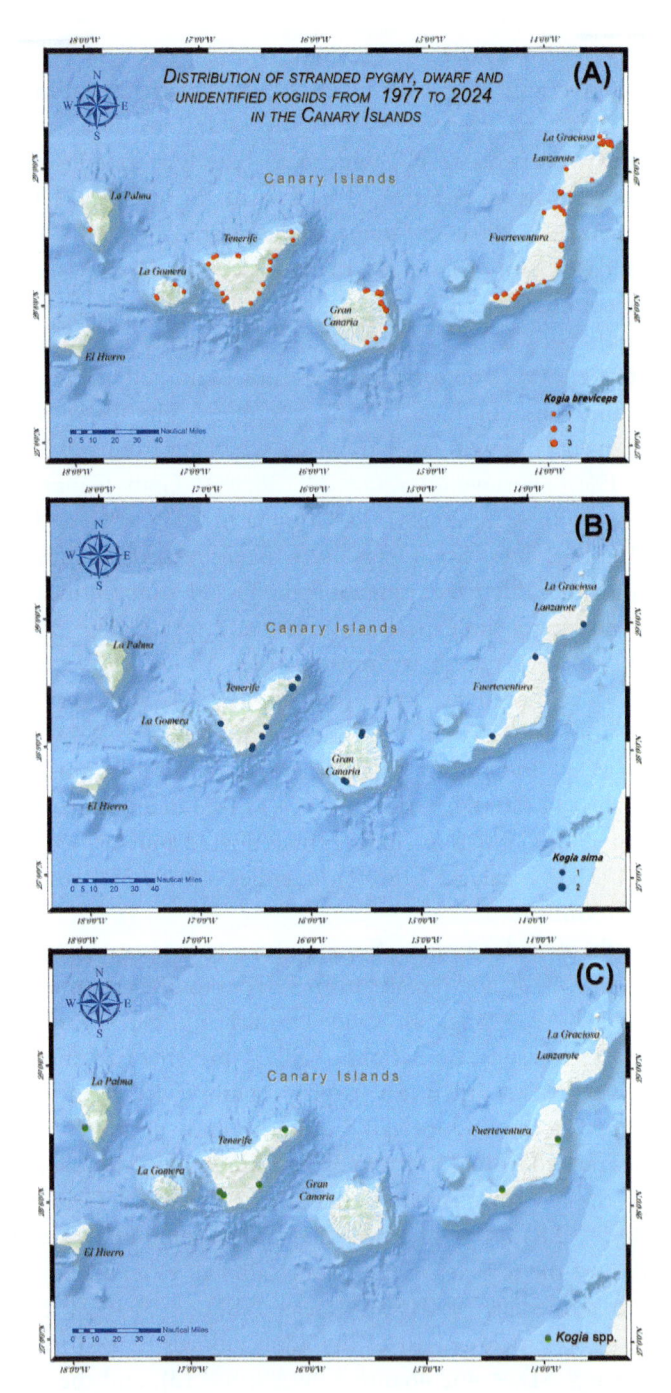

Fig. 2 Distribution of stranded pygmy sperm whale, *K. breviceps* (A), dwarf sperm whale, *K. sima* (B), and unidentified kogiids (C) from 1977 to 2024 in the Canary Islands.

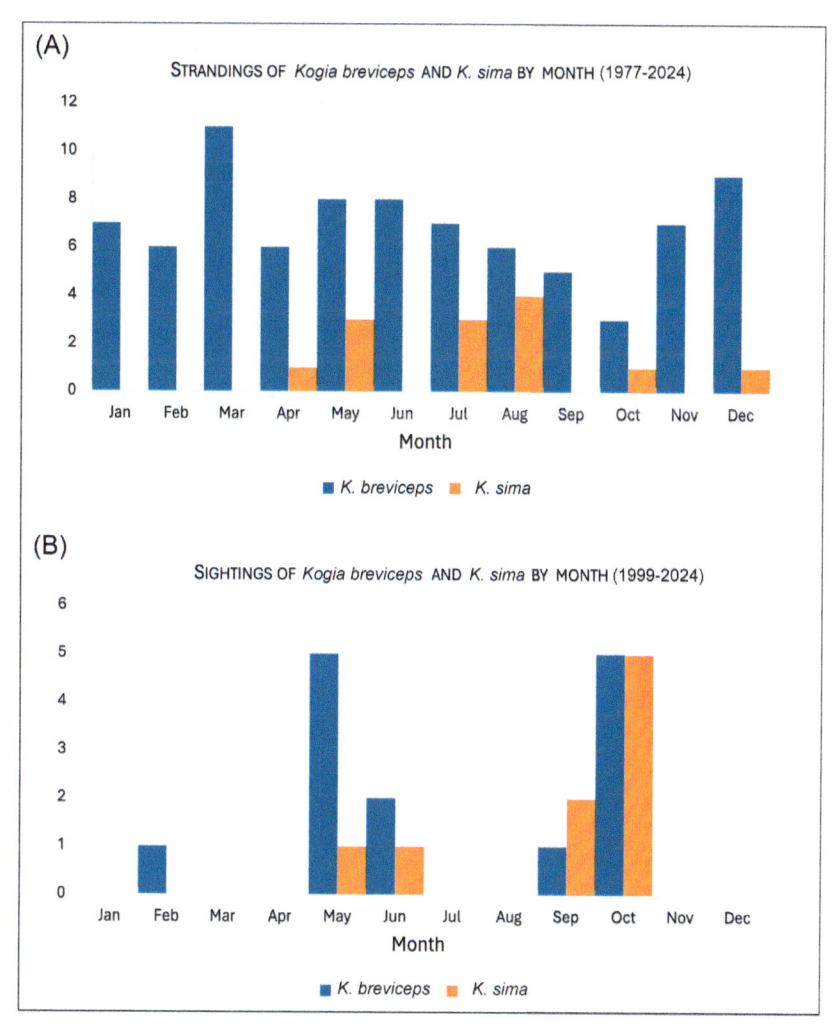

Fig. 3 Histogram of *K. breviceps* and *K. sima* strandings by month in the Canary Archipelago during the study period (A). Histogram of *K. breviceps* and *K. sima* sightings during the study period (B).

(mean ± SD = 452.62 mm ± 65.56 mm; $n = 13$). The testicles of adult males had well-developed epididymides and vas deferens, usually with abundant seminal fluid.

A 225 cm pregnant female with a 10 cm fetus (Kbr22), whose length is considerably below the length of sexual maturity in females of 262 cm given by Plön (2004), suggests a potential error in measuring the total length of this specimen. The shortest mature females in the sample are a

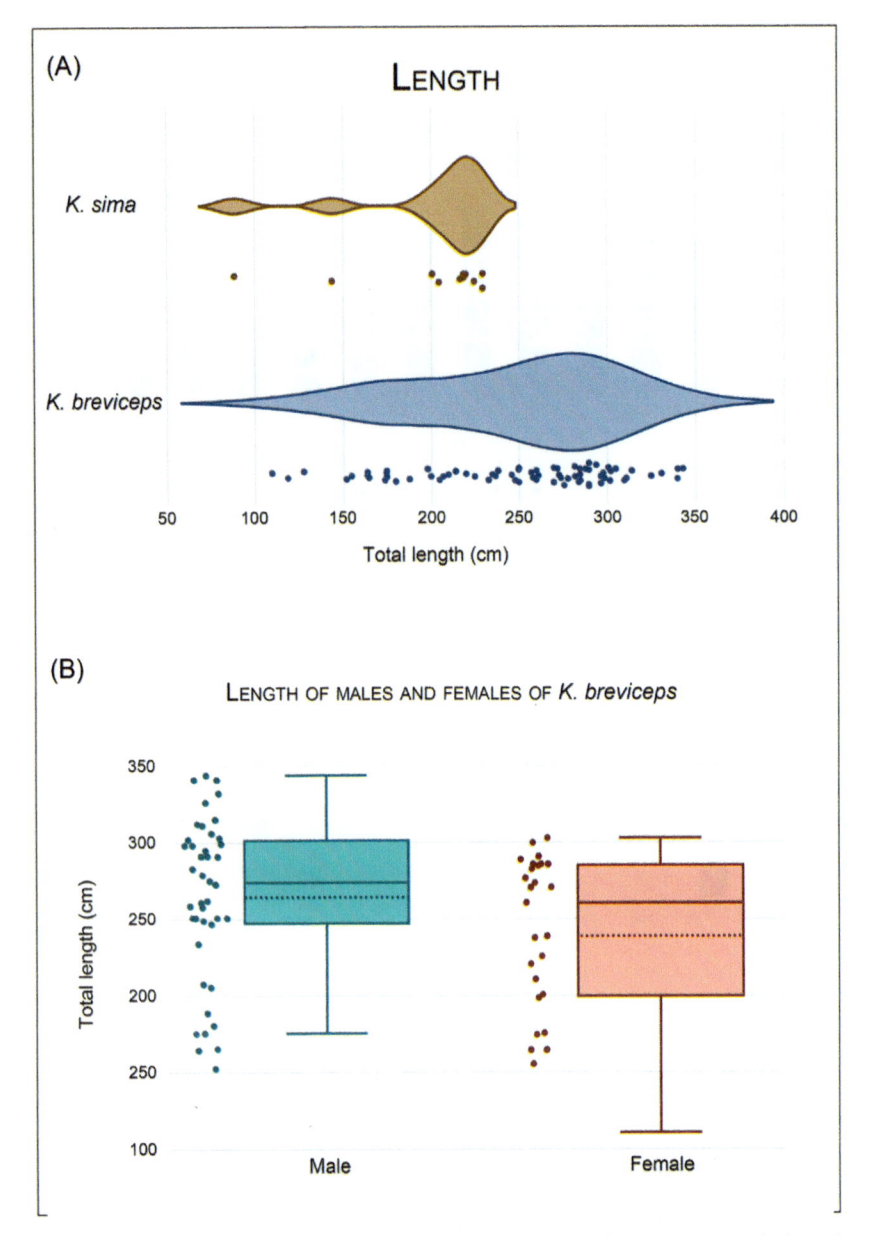

Fig. 4 Violin plot showing the lengths of *K. sima* and *K. breviceps* stranded in the Canary Archipelago during the study period (A), box plot displaying the lengths of male pygmy sperm whales stranded in the Canary Islands during the study period (B). The solid line represents the mean, and the dashed line represents the median.

Table 1 Minimum and maximum weights recorded for pygmy sperm whales of both sexes stranded in the Canary Archipelago during the study period.

Data	Sex	LT (cm)	Weight (kg)	Code	Island
November 24, 2020	Male	301	360	Kbr75	TF
April 17, 2006	Female	290	420	Kbr38	TF
July 26, 2012	Female	110	19	Kbr55	TF
March 20, 2014	Female	155	72	Kbr60	TF
January 29, 2018	Female	164	56	Kbr70	TF
April 27, 2010	Male	165	60	Kbr47	TF
January 16, 2018	Male	175	72	Kbr69	FV

260 cm pregnant female with an 18 cm fetus (Kbr42), stranded in Lanzarote on August 29, 2007, and a 270 cm pregnant female (Kbr03) with a dependent calf stranded in Haría, Lanzarote on November 25, 1988. There were four cases of mother-calf stranding events. Three females measuring 270 cm, 273 cm, and 284 cm were accompanied by calves measuring 175 cm, 152 cm, and 164 cm in length, respectively. A calf of 155 cm long was stranded and associated with an adult-sized animal that was refloated and presumed to be its mother. The first two females were pregnant, with fetuses measuring 11 cm and 4 cm long, respectively.

The length of pregnant females ranged from 260 cm (pregnant with a fetus of 18 cm in length) to 299 cm (median= 279, mean ± SD = 273.7 ± 21.7; $n = 9$). The length of the fetuses ranged from 4 cm to 63 cm long (mean ± SD = 23.2 ± 19.2; $n = 9$). Based on the linear part of the fetal growth curve used by Kasuya's (1977) equation, Plön (2004) calculated a fetal growth rate of 0.34 cm/day for *K. breviceps* and 0.31 cm/day for *K. sima*. Considering this, the fetuses of *K. breviceps* from the Canary Islands sample were between 11.8 and 105.3 days old, with a median of 68.3 days (mean ± SD = 68.3 ± 56.51; $n = 9$). Mother-calf pairs represent 44.4% of the pregnant females and 23.5% of the mature females in the sample. Neonates measuring 110 cm (Kbr55), 119 cm (Kbr48) (still with umbilical cord), and 128 cm (Kbr81) stranded in the Canary Islands, with the first two in July and the third in October. Individually stranded calves, considered suckling based on their total length, and four calves accompanied by their mothers, ranged from 152 to 175 cm, with a median of 165 cm (mean ± SD = 165.71 ± 9.55; $n = 7$) had solid food in their stomachs. Table 2 presents the calves and juvenile pygmy sperm whale specimens of both sexes stranded in the Canary Archipelago during the study period, all of which had solid food in their stomachs (Figs. 5 and 6).

3.1.2 Dwarf sperm whale

For the dwarf sperm whale, stranding events were as follows: 2 events (14.3%) with 3 animals were live stranded (active), 9 events (64.3%) with 9 animals were dead (passive), and 3 events (21.4%) with 3 animals were indeterminate. Two live-stranded animals were returned to the sea (including a 220 cm female). For dead dwarf sperm whales, two carcasses were not examined, and postmortem studies were carried out on 11 specimens. Stranding events of *Kogia sima* by island were as follows: Lanzarote and La Graciosa ($n = 1$; 7.1%), Fuerteventura ($n = 2$; 14.3%), Gran Canaria ($n = 4$; 28.6%), and Tenerife ($n = 7$; 50.0%) (Fig. 2B). Except for two in October and December, strandings have been concentrated from April to August (Fig. 3A). Carcass condition included 5 (45.5%) very fresh and fresh animals, 2 (18.2%) with moderate autolysis, 1 (9.1%) with advanced autolysis, and 3 (27.3%) with very advanced autolysis. Dwarf sperm whale strandings constituted 0.91% of the total cetacean stranding events in the Canary Islands between 1977 and May 2024, representing the twelfth most frequent species in stranding events.

The sex distribution for all individuals was 9 (60.0%) females, 2 (13.3%) males, and 4 (26.7%) indeterminate. The age categories for all individuals were 2 (13.3%) neonate/calves, 1 (6.7%) juvenile/subadult, 6 (40.0%) adults, and 6 (40.0%) indeterminate. The lengths of the two mature males of the sample were 225 cm and 230 cm, respectively. The length of females ranged from 80 cm to 230 cm (mean ± SD = 190.9 ± 49.1; $n = 8$), and the length of mature females ranged from 201 cm to 230 cm (mean ± SD = 214.8 ± 11.8; $n = 5$; Fig. 4). There was one record of a mother-calf pair stranding on July 1, 1995, in Santa Cruz de Tenerife. The calf was refloated, and the 219 cm pregnant female (Ksi02), presumed to be its mother, had a fetus measuring 92.5 cm long. According to Plön's (2004) calculations of a fetal growth rate of 0.31 cm/day for *K. sima*, the 92.5 cm fetus would be approximately 298.4 days old. Two mature non-pregnant females of 201 cm (Ksi07) and 230 cm (Ksi14), stranded on May 23, 2006 and August 23, 2018, respectively, had gestational *corpora luteum* in the left ovary. In the case of Ksi14, the left uterine horn was dilated, and milk was observed in the mammary gland, suggesting a recent gestation.

Table 2 Calves and juvenile pygmy sperm whale specimens of both sexes stranded in the Canary Archipelago during the study period, all of which had solid food in their stomachs.

Data	Sex	LT (cm)	Code	Observations
March 28, 2011	Male	175	Kbr50	Cephalopod sucker marks on the facial region
March 20, 2014	Female	155	Kbr60	Tunicates, cephalopod beaks, and plastics in its stomach
December 8, 2014	Female	174	Kbr63	All teeth were beginning to emerge
November 3, 2023	Female	164	Kbr86	Some teeth erupted while gums covered others

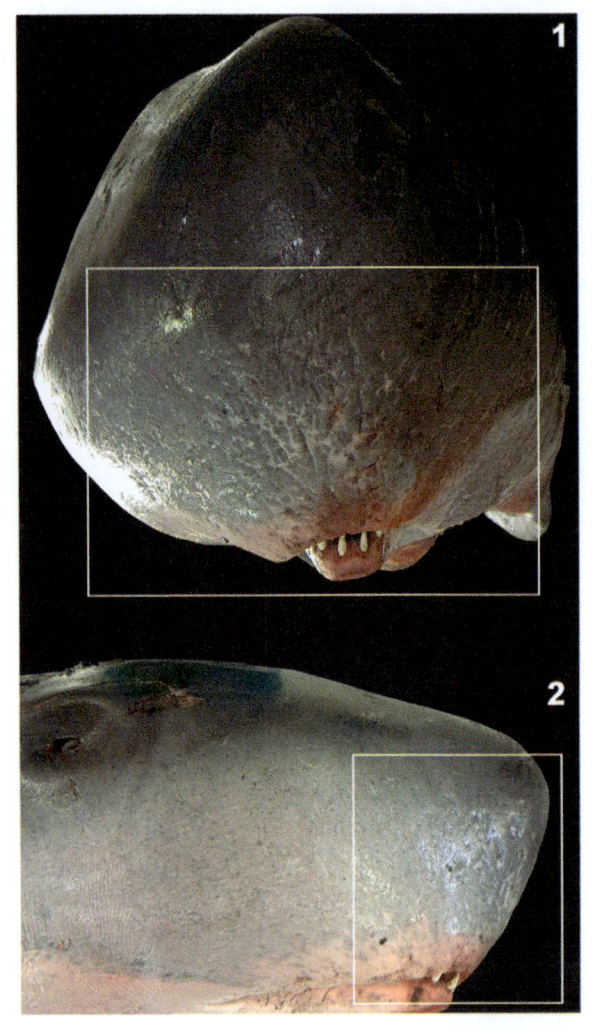

Fig. 5 Frontal and lateral views of the head of an adult male pygmy sperm whale (*K. breviceps*) stranded in Tenerife Islands on March 15, 2023, with extensive old abrasion and scarring in the front of snout, suggestive of bottom foraging.

A freshly stranded 302 cm adult female (Kbr77) and a 305 cm adult male pygmy sperm whale (Kbr82) exhibited extensive abrasions and scarring on their snouts that were old and not caused in the stranding moment (Fig. 5). Stranded specimens of both species exhibited small antemortem scars from old bites and wounds inflicted by unidentified small sharks, particularly on the flippers, dorsal fin, and flukes, but did not show healed or fresh cookiecutter shark (*Isistius brasilensis*) or largetooth cookiecutter shark (*I. plutodus*) bites.

Strandings and sightings at sea 37

Fig. 6 Dwarf sperm whale juvenile stranded alive on Tenerife Island on May 2, 2001 (1). Pygmy sperm whale calf stranded on Gran Canaria Island on April 3, 2014 (2). Head detail of the pygmy sperm whale calf stranded on Gran Canaria Island on April 3, 2014 (3). Dwarf sperm whale mature female stranded in Majanicho, Fuerteventura Island on February 24, 2014 (4). A mature female, 260 cm in length, was found floating off Papagayo on August 29, 2007, with scars from interspecific interaction with killer whales (5).

A high portion of the carcasses showed postmortem bites from sharks. On May 20, 2016, a 219 cm female dwarf sperm whale (Ksi12) was found stranded in Las Eras, Fasnia, Tenerife. The animal exhibited a 5 cm linear and penetrating lesion located between the left eye and the blowhole, extending to the skull, which presented cranial trauma. The characteristics of the lesion could be compatible with an injury caused by the bill of a swordfish. Twelve animals (ten *K. breviceps* and two *K. sima*) exhibited signs of interspecific aggressive interactions or predation by killer whales (*Orcinus orca*), including acute cutaneous punctate wounds and tooth rake marks on the skin and blubber, with intertooth spacing consistent with that of killer whales. Of these, 8 (66.6%) were females, of which 5 (62.5%) were pregnant, and three were males (two juveniles and one of undetermined age; Figs. 6 and 7; Table 3). A forensic necropsy revealed that a 305 cm long adult male *K. breviceps* (Kbr82), which stranded in Tenerife on March 15, 2023, with recent marks from an orca attack, survived for a period before its death (Fig. 7).

3.2 Sightings

Between 1999 and 2024, we conducted 1438 days of dedicated cetacean surveys in the Canary Islands, covering 13,562 km and 12,074 h of survey effort, resulting in a total of 7863 cetacean sightings (Fig. 8A). We recorded 36 sightings of kogiids in the Canary Islands, 34 during fieldwork in dedicated surveys, and 2 in opportunistic sightings from citizen science. Of these, 14 (39%) were *K. breviceps*, 9 (25%) were *K. sima*, and 13 (36%) were unidentified *Kogia* (Fig. 8B). Table 4 presents the survey effort, total sightings, and sightings of *K. breviceps*, *K. sima* and unidentified *Kogia* in the waters off the eastern coasts of Lanzarote and Fuerteventura, as well as in the remaining regions of the Canary Islands. Twenty-nine sightings (80.5%) of kogiids were recorded in the waters off the eastern coast of the islands of Lanzarote and Fuerteventura: 11 (37.9%) of pygmy sperm whale, 8 (27.5%) of dwarf sperm whale, and 10 (34.4%) of unidentified kogiids (Fig. 9A).

3.2.1 Pygmy sperm whale

In pygmy sperm whales, the total time spent with the animals was 218 min, with encounter duration ranging from less than one minute to 84 min (mean ± SD = 16.8 ± 25.3; $n = 13$). We obtained 523 photographs of pygmy sperm whales and 748 photographs of dwarf sperm whales. In pygmy sperm whales, the sightings occurred in the following sea conditions: Beaufort 0 ($n = 9$; 64.3%) and Beaufort 1 ($n = 5$; 35.7%). Pygmy

Strandings and sightings at sea 39

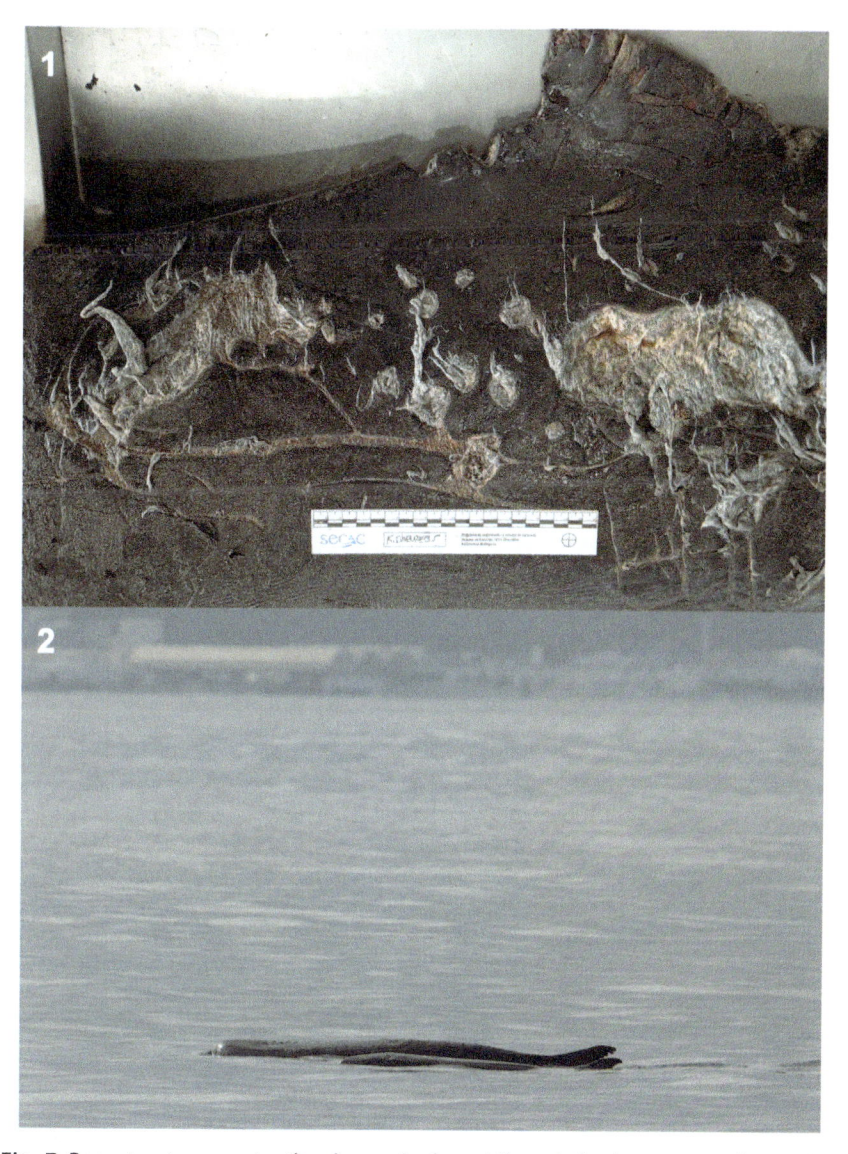

Fig. 7 Recent cutaneous tooth rake marks from killer whales in a 305 cm long adult male *K. breviceps* stranded in Tenerife on March 15, 2023, which survived for a time before its death (1). Female pygmy sperm whale accompanied by a calf, logging off Arrecife, Lanzarote on May 27, 2016 (2).

sperm whales were sighted at water depths ranging from 618 m to 1815 m (mean ± SD= 1204.3 ± 397.75; $n = 13$). The animals were recorded at a distance from the coast ranging from 4746 m to 28,270 m (mean ± SD =

Table 3 Published and unpublished pygmy and dwarf sperm whales stranded in the Canary Islands displayed signs of interspecific aggressive interactions or predation by killer whales (*Orcinus orca*), such as acute cutaneous punctate wounds and tooth rake marks on the skin and blubber.

Data	Sex	LT (cm)	Code	Class	Island	Observations
September 25, 2005	Female	299	Kbr33	Adult	GC	Pregnant
February 2, 2006	Male	205	Kbr35	Young	GC	
March 31, 2006	Female	288	Kbr37	Adult	LG	Pregnant*
August 13, 2006	Female	285	Kbr39	Adult	FV	
April 6, 2007	Female	282	Kbr40	Adult	TF	Pregnant*
June 20, 2007	Ind.	Ind.	Kbr41	Ind.	GC	
August 28, 2007	Female	260	Kbr42	Adult	LZ	Pregnant*
December 14, 2014	Female	174	Kbr63	Young	FV	
August 23, 2018	Female	230	Ksi14	Adult	LZ	Pregnant
September 9, 2018	Male	225	Ksi15	Adult	TF	
March 17, 2023	Male	305	Kbr82	Adult	TF	
September 19, 2023	Female	270	Kbr84	Adult	GC	

GC (Gran Canaria), LG (La Gomera), TN (Tenerife), FV (Fuerteventura) and LZ (Lanzarote). *Lozano et al. (2020).

Strandings and sightings at sea 41

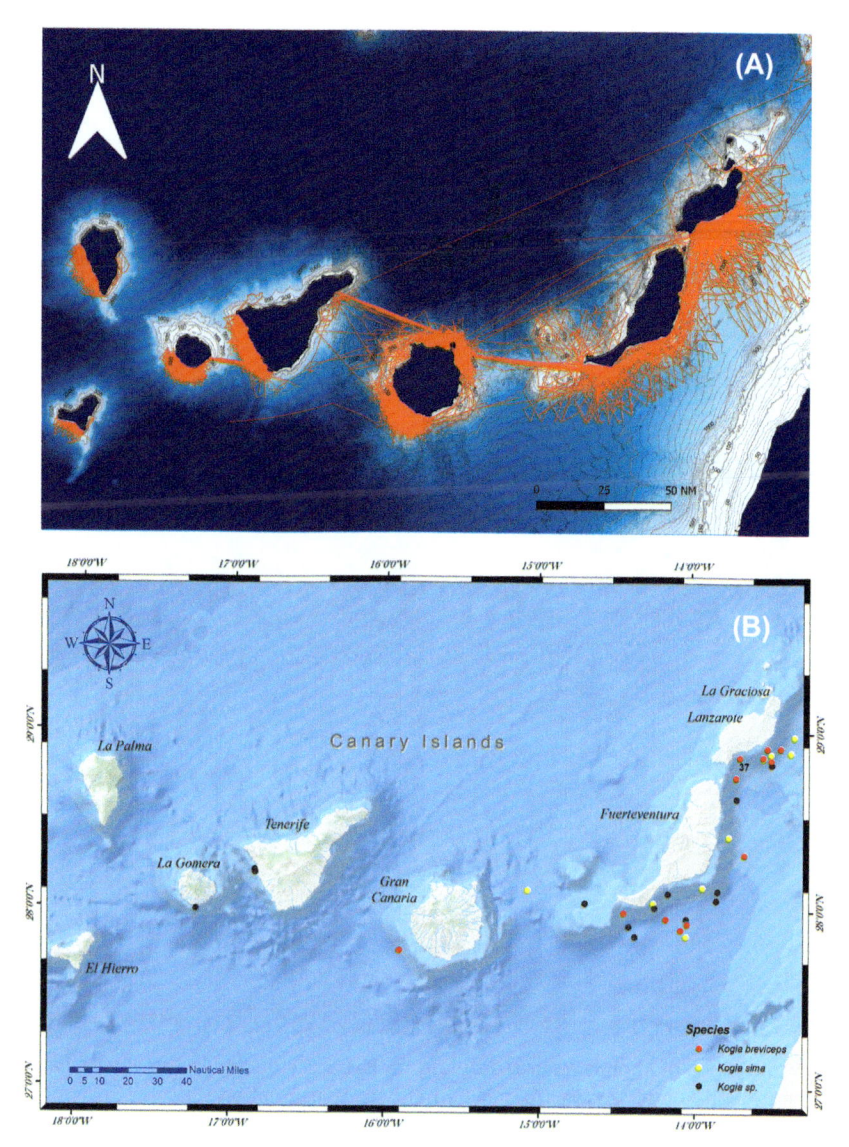

Fig. 8 Transects conducted in the waters of the Canary Islands during dedicated cetacean surveys in the study period (A), and the distribution of at-sea sightings of pygmy sperm whales (*K. breviceps*), dwarf sperm whales (*K. sima*), and unidentified kogiids during this period (B).

12,882.9 ± 7988.8; $n = 13$). Group sizes of pygmy sperm whales ranged from 1 to 2 animals (Mean ± SD = 1.46 ± 0.5; $n = 13$). The presence of calves was detected on 4 (30.8%) occasions for *K. breviceps* (Fig. 7).

Table 4 Survey effort, total sightings, and sightings of *Kogia* in the waters off the eastern coasts of Lanzarote and Fuerteventura, as well as in the remaining regions of the Canary Islands.

Survey effort	Eastern Lanzarote and Fuerteventura	Remaining regions of the Canary Islands
Days	421	1017
Km	3936	9626
Hours	3712	8362
Sightings	1597	6266

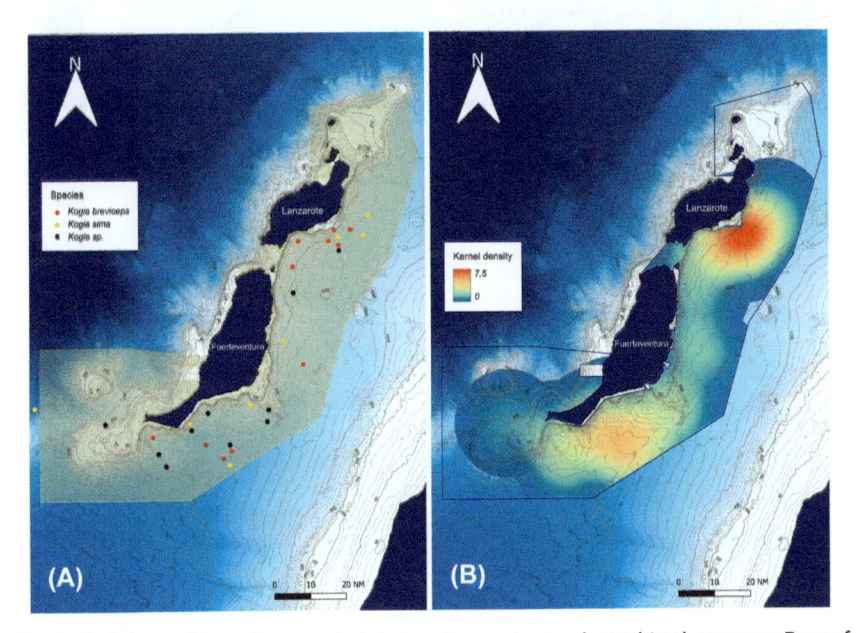

Fig. 9 Sightings of kogiids recorded during transects conducted in the waters East of Lanzarote and Fuerteventura (A) and Kernel density estimation (B).

3.2.2 Dwarf sperm whale

For dwarf sperm whales, the total time spent with the animals was 160 min, ranging from less than one minute to 37 min (mean ± SD = 17.8 ± 13.5; $n = 9$). We obtained 748 photographs of dwarf sperm whales. A group of 3 *K. sima* observed on September 24, 2019, off the coast of Lanzarote was recorded with an unmanned aerial system (drone; Fig. 10). In dwarf sperm whales, the sightings occurred in the following sea conditions: Beaufort 0

Fig. 10 Pygmy sperm whale logging off SE Lanzarote on September 19, 2022 (1). Mother-calf pairs of dwarf sperm whale logging off SE Fuerteventura on October 18, 2010 (2). An individual from a group of 3 dwarf sperm whales observed on September 24, 2019, off the coast of Lanzarote was recorded with an unmanned aerial system (UAS) (3). Potential dependent calf of a pygmy sperm whale, with an estimated length of less than 2 m, remaining alone at the surface throughout the sighting, observed on October 1, 2000 (4).

(n = 5; 55.5%) and Beaufort 1 (n = 4; 44.4%). Dwarf sperm whales were sighted at water depths ranging from 930 m to 2208 m (mean ± SD = 1323.8 ± 423.9; n = 9). The dwarf sperm whale was found at a distance from the coast ranging from 3582 m to 32,827 m (mean ± SD = 14,922 ± 10,506.2; n = 9). Group sizes of dwarf sperm whales ranged from 1 to 3 individuals (mean ± SD = 1.78 ± 0.83; n = 9). The presence of calves was detected on 3 (33.3%) occasions for *K. sima* (Fig. 10.2).

3.2.3 Unidentified Kogia

The total time spent with unidentified *Kogia* was 138 min, ranging from less than one minute to 33 min (mean ± SD: 10.6 ± 12.4; n = 13). In unidentified *Kogia* the sightings occurred in the following sea conditions: Beaufort 0 (n = 6; 46.1%), Beaufort 2 (n = 1; 7.7%), and Beaufort > 3 (n = 1; 7.7%). The sighting depth of encounters with unidentified *Kogia* ranged from 380 m to 1797 m (mean ± SD = 1314.4 ± 414.8; n = 10). The distance from the coast in encounters with unidentified *Kogia* ranged from 588 m to 25,710 m (mean ± SD = 12,288.2 ± 8203.2; n = 13). Group sizes of unidentified *Kogia* spp. ranged from 1 to 2 animals (Mean ± SD = 1.46 ± 0.52; n = 13). The presence of calves was detected on only one occasion. Except for a pygmy sperm whale sighting in February, sightings of both species in the archipelago at sea were recorded from May to June and from September to October (Fig. 3B).

There was no significant difference in the depths of encounters with pygmy and dwarf sperm whales (Mann-Whitney U-test U = 48; p = 0.512; r = 0.15). The median depth of encounters with kogiid calves (including unidentified *Kogia*) was 1012.5 m (mean = 1055.31; SD = 372.47; n = 7), and the median depth of groups without calves was 1299.9 m (mean = 1420.6 m; SD = 353.57; n = 19; Fig. 11A). The median distance from the coast of groups with calves was 7497.2 m (mean = 9440.1; SD = 8538; n = 7), and the median distance of groups without calves was 14,003.8 m (mean = 16,076.3 m; SD = 8235.4 m; n = 19; Fig. 11B). *Kogia* groups with calves were significantly shallower (Mann-Whitney U-test U = 26, p = 0.018, r = 0.46) and closer to the coast (Mann-Whitney U-test U = 25, p = 0.015, r = 0.47) than *Kogia* encounters without calves. In the sightings of kogiids in the waters off the eastern coast of the islands of Lanzarote and Fuerteventura, the Kernel density analysis reveals two core areas: one off the central coast of Lanzarote and another in the south of Fuerteventura (Fig. 9B). Most sightings were recorded outside the slopes of both islands.

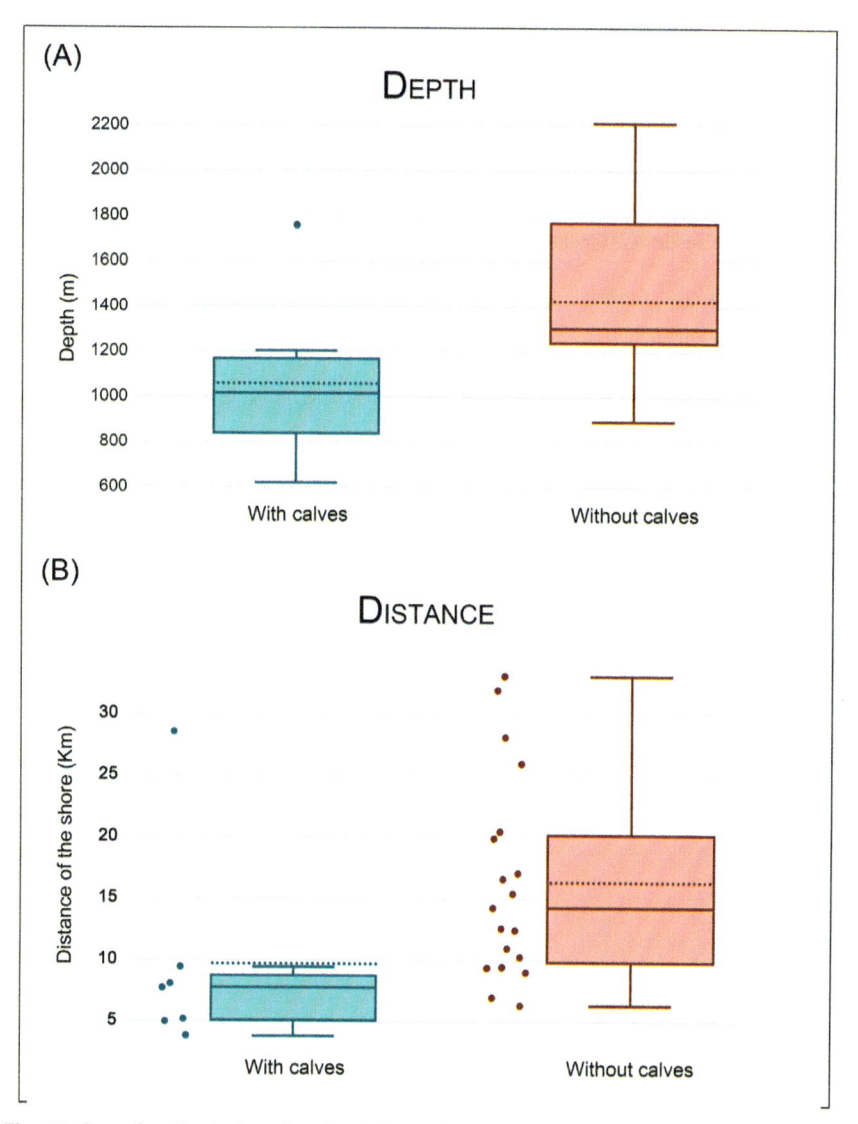

Fig. 11 Box plot displaying the depth (A) and the distance from the shore (B) of *Kogia* sightings, with groups having calves and without calves in the Canary Islands. The solid line represents the mean, and the dashed line represents the median.

In pygmy sperm whale sightings calves were always in mother-calf pairs, with no other individuals associated. On October 1, 2000, we observed a potentially dependent calf of *K. breviceps* with an estimated length of less than 2 m. The calf remained at the surface throughout the sighting, actively moving and staying away in response to the vessel's presence until our departure from the area. Several minutes later, at the same location, we observed a mother-calf pair in the distance (Fig. 10.4). On October 4, 2009, during an encounter off Fuerteventura's eastern coast, we observed two adult-sized individuals separated by several meters, one located behind the other. Each exhibited distinct morphological characteristics that allowed us to identify them as *K. sima* (Fig. 12.1) and *K. breviceps* (Fig. 12.2), respectively. On one occasion, during which the animals were observed actively moving, the initial behavior of kogiid whales was logging, lying motionless at the surface. In two sightings of mother-calf pairs of dwarf sperm whales with large-sized calves, we observed individuals distancing themselves by several meters after the dive, remaining motionless at the surface upon resurfacing (Fig. 13). On no occasion was breaching or aerial behavior observed. However, during several transect surveys, we observed numerous isolated breaches in the distance by unidentified medium-sized cetaceans, which could have been either beaked whales or *Kogia* spp. species. Two encounters with pygmy sperm whales were associated with the presence of Gervais' beaked whale (*Mesoplodon europaeus*) within a 1000-meter radius around the vessel. In two separate sightings of dwarf sperm whales, Cuvier's

Fig. 12 Two apparent adult sized Kogia individuals separated by several meters were observed on October 4, 2009, off the eastern coast of Fuerteventura, each exhibiting distinct morphological characteristics and identified as a dwarf sperm whale (1) and a pygmy sperm whale (2), potentially representing the first documented record, to our knowledge, of a spatially-associated sighting of both species.

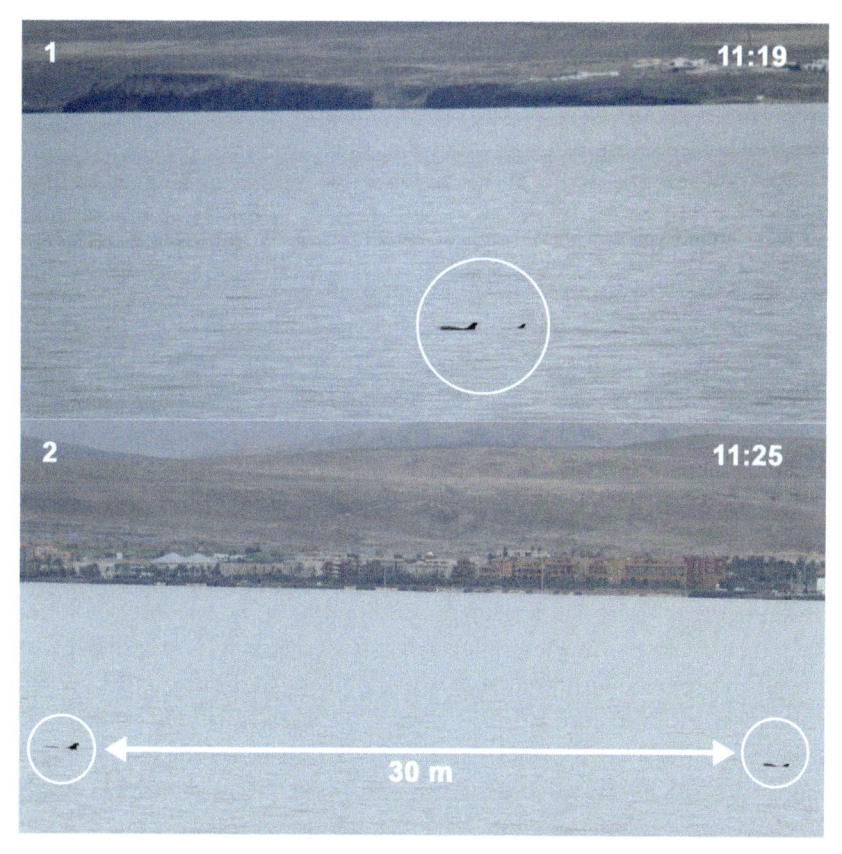

Fig. 13 Mother-calf pairs of dwarf sperm whales were observed off the Fuerteventura coast on October 18, 2010. Both individuals distance themselves by several meters after a 9 min dive, remaining motionless at the surface upon resurfacing.

beaked whale (*Ziphius cavirostris*) and Blainville's beaked whale (*M. densirostris*) were observed within the same area. Additionally, during two encounters with unidentified kogiids, Gervais' beaked whale and Atlantic spotted dolphins (*Stenella frontalis*) were also present. In all cases, no interspecific interaction or association with the groups of other species was observed.

4. Discussion
4.1 Strandings

The high number of kogiid strandings in the region reveals that the Canary Archipelago is a frequent hotspot for strandings of both species, especially the pygmy sperm whale, similar to other regions, such as South Africa and the southeastern United States (Plön, 2004; Hodge et al., 2018; Odell, 1991). Although the Canary Islands have a substantially shorter coastline compared to these regions, the stranding rates are high, particularly considering the low probability of drifting carcasses reaching the shore due to: (i) the Canary Islands' coasts facing open waters, (ii) the existence of windy inter-island channels that increase the likelihood of carcass drift southward and reduce the probability of stranding, and (iii) long stretches of rocky and inaccessible coastline, which could result in a high potential for carcass loss and unreported strandings. As a result, the average annual stranding record of cetaceans in general in the Canary Archipelago is significantly lower than that of regions with a similarly long coastline (M. Sequeira and A. López, *personal communication*).

It is unclear which environmental drivers may contribute to the observed *Kogia* stranding patterns in the Canary Islands. Moura et al. (2016) found that the most important environmental predictor explaining stranding events along the Brazilian coastline was wind speed, which may have a significant postmortem effect on carcass drift and on the debilitated or health-compromised vector of the population. Berini et al. (2015) identified significant correlations between *K. breviceps* strandings and the monthly multivariate El Niño Southern Oscillation Index, sea surface temperature (SST), wind speed (with more strandings occurring during periods of high wind speeds), wave height, average wave period, and barometric pressure for strandings in the southeastern USA. In the Canary Archipelago, there is a difference between the proportion of stranded animals and sightings of both species, which is significantly more evident in strandings, with pygmy sperm whales more frequently recorded as stranded than dwarf sperm whales. However, it is important to interpret stranding records carefully when considering strandings versus sightings. For example, in Hawai'i, dwarf sperm whale strandings are only about one-fourth as frequent as those of pygmy sperm whales, yet dwarf sperm whales represent over 90% of *Kogia* sightings around the central Hawaiian Islands (Baird et al., 2021).

The maximum lengths for pygmy sperm whales in South Africa were reported as 330 cm for males and 327 cm for females (Plön, 2004). Two male pygmy sperm whales stranded in the Canary Islands exceeded these lengths: one measuring 343 cm stranded in Arucas, Gran Canaria, on June 13, 2015, and another measuring 340 cm stranded in Hermigua, La Gomera, on January 3, 2012. The maximum recorded weights for the species include a 317 cm individual weighing 450 kg in Canada (McAlpine et al., 1997) and two females from South Africa weighing 480 kg each (Plön, 2004).

4.2 Reproduction

The estimated lengths at birth for *K. breviceps* and *K. sima* are 120 cm and 103 cm, respectively (Plön, 2004; Best, 2007). The lengths at sexual maturity for males are 241–242 cm for *K. breviceps* and 197 cm for *K. sima*, and for females, they are 262 cm and 215 cm, respectively (Plön, 2004). A 201 cm adult female dwarf sperm whale stranded in Majanicho, Fuerteventura Island, on February 24, 2014, with a regressing gestational *corpus luteum*, was slightly below the 215 cm length at sexual maturity reported for females by Plön (2004). Considering a length at birth of 120 cm, a gestation period of 11.8 months, and a growth rate of 0.34 cm/day derived from Kasuya's (1977) equation (Plön 2004), we estimate that conception in pygmy sperm whale in the Canaries occurs between June and February, with a peak between September and December. Births occur from May to January, peaking between August and November. This birth period aligns with the months of three neonate stranded in the Canary Islands. The highest average sea surface temperatures in the Canary Archipelago are recorded from August to October, at around 22–23 °C. These conditions may present more optimal thermal environments for neonatal development. Insufficient information is available on the dwarf sperm whale.

Female pygmy and dwarf sperm whales can have a postpartum oestrus, meaning that lactating females are able to conceive shortly after giving birth (Plön, 2004). In the Canary Archipelago, four cases of mother-calf pairs of *K. breviceps* were documented, two of which involved pregnant females. However, this number could be higher due to sampling artifacts and the potential separation of mother-calf pairs at sea prior to stranding for various reasons. In South Africa, the percentage of simultaneously lactating and pregnant females was 24.1% in pygmy sperm whales and 11.5% in dwarf sperm whales (Plön, 2004), and the largest calf of *K. breviceps* stranded with a lactating female measured 202 cm (Plön, 2004). Pygmy sperm whale weaning appears to occur at lengths between 171 cm and 202 cm, and weaning may happen between the first and second year. The close association between mothers and calves likely ends shortly after weaning (Ross, 1979; Plön, 2004). Ross (1979) notes that calves can begin ingesting solid food at lengths of 160 cm, and examinations of stomach contents from lactating calves in the Canary Islands support this assertion (Table 2).

4.3 Presence in the Canary Archipelago

The spatiotemporal patterns of strandings in the Canary Islands indicate that pygmy sperm whales are present year-round in this archipelago. In contrast, the majority of dwarf sperm whale strandings occurred from April to December, with a peak from April to August (Fig. 3A). Additionally, pygmy sperm whale strandings occurred across all the islands (Fig. 2A), while most dwarf sperm whale strandings occurred in the central islands of the archipelago (Fig. 2B). In contrast, most sightings of *Kogia* spp. at sea in the archipelago occur in May, September, and October (Fig. 3B), which are the months with the highest number of calm days reflecting the ability to visually detect *Kogia* species at sea (Barlow, 2015). The seasonal distribution of strandings and at sea observations suggests a year-round presence of pygmy sperm whales, while dwarf sperm whales might have a more seasonal presence in the islands, where they seem to coexist sympatrically in the same habitats during certain months.

A significant number of sightings of kogiids in the Canary Archipelago were recorded in the waters off the eastern coast of the islands of Lanzarote and Fuerteventura (Fig. 9A), despite considerably less effort being made in this region compared to other areas of the Canary Islands (Table 4), suggesting that this sector may be an important area for both species within the archipelago. The Canary Islands are situated in an oceanographic transition zone between the cold, nutrient-rich upwelling system off the northwestern African coast and the warmer, oligotrophic open waters (Barton et al., 1998). Filaments from this upwelling reach the islands from the African continental shelf, contributing to transporting organic matter such as fish, cephalopods, and crustacean larvae. These water masses with high chlorophyll and primary production decrease towards the ocean and exhibit spatial and seasonal variations driven by the trade winds along northwest Africa (Ariza et al., 2016). Mesoscale oceanographic phenomena such as the 'island-mass effect' (Doty and Oguri, 1956) and 'island stirring' (Mann and Lazier, 1991) create wakes and a complex system of island-generated eddies, processes that significantly contribute to local productivity by acting as nutrient pumps (Aristegui et al., 1994; Ariza et al., 2016). The waters off the eastern coast of the islands of Lanzarote and Fuerteventura represent a differentiated marine habitat from the rest of the Canary Islands due to their proximity to the neighboring African coast. The 'Lanzarote Passage' is a channel of relatively shallow water (up to 1300 m) between the island of Lanzarote and Fuerteventura and the

African continental shore (Hernández-Guerra et al., 2003). Between the island slopes and the African continental margin, there is a wide abyssal plain. To the north and south of this passage, the depth of the seafloor increases sharply (Fig. 8A) and its geomorphological characteristics and oceanography favor high biological nutrient enrichment that increases productivity. These factors locally enhance the productivity of the region compared to the rest of the archipelago and support a complex and diverse mesopelagic food web, constituted by a diverse and abundant assemblage of fishes, cephalopods and crustaceans (Badcock, 1970; Bordes et al., 1999, 2009) and serve as the primary trophic source for open-ocean predators such as deep-diving cetaceans (Braun et al., 2022). Some encounters with kogiids off Lanzarote and Fuerteventura occurred during a study on beaked whales, suggesting that these species share a habitat in this area.

The pygmy and dwarf sperm whales are deep-diving cetaceans that predominantly feed on mesopelagic and bathypelagic cephalopods (especially Histioteuthidae and Cranchidae), fish, and crustaceans, which they capture within the water column and near the ocean floor (West et al., 2009). Both species share a similar diet (Plön, 2004; Staudinger et al., 2014). Stomach content, isotopic, gut microbiome, and persistent organic pollutant (POP) analyses indicate a high niche overlap between the two (Plön, 2004; Erdwin et al., 2017; Plön, 2022; Plön and Baird, 2022; De Oliveira Ferreira et al., 2023; Tasch et al., 2023). However, *K. sima* appears to have a more specialized diet compared to *K. breviceps*, with records of 15 families of cephalopods and 16 families of fish (Plön, 2004; Staudinger et al., 2014). Additionally, *K. sima* consumes slightly smaller prey than *K. breviceps* (Plön, 2004; Staudinger et al., 2014). Cephalopods are a crucial component of marine food webs. The Canary Archipelago is a hotspot presenting a high diversity of species (Escánez et al., 2022) and the primary resource in the diet of deep-diving cetaceans in the Canary Islands (Hernández-García and Martín, 1996; Santos et al., 2007). In the stomach contents of 7 pygmy sperm whales stranded in the Canary Archipelago between 1987 and 1993, remains of unidentified Trichiuridae and Myctophidae fish, *Lepidopus caudatus*, and the cephalopods *Tadarodes sagittatus*, *Histioteuthis* (type A), *H. bonnelli*, *H. melagroteuthis*, *Cycloteuthis* sp., *Taonius* sp., *Teuthowenia megalops*, *Vampiroteuthis* sp. and *Pholidoteuthis adami*, as well as unidentified crustaceans (Hernández-García and Martín, 1996) were identified. Two specimens stranded in Tenerife (a 245 cm male, and a 188 cm male, stranded on 1 May 2002 and 21 June 2006, respectively) had the following prey species in their stomach contents: Ommastrephidae (*Taradores sagittatus*), Sepiolidae (*Steloteuthis* sp.), Architeuthidae (*Architeuthis* sp.), Vampiroteuthidae

(*Vampiroteuthis infernalis*), Cranchidae (*Megalocranchia* sp., *Cranchia* sp., *Lyocranchia* sp., and *Taonius pavo*), Mastigoteuthidae (*Mastigoteuthis* sp.), Histioteuthidae (*Histioteuthis meleagroteuthis*, *H. reversa*, and *H.* type A), Pholidoteuthidae (*Pholidoteuthis bochmai*), Chiroteuthidae (*Chiroteuthis veranyi* and C. type II), Onychoteuthidae (*Moroteuthis* sp.), and Octopoteuthidae (*Octopoteuthis* sp.), and fish prey Trichiuridae (*Aphanopus carbo*), Myctophidae (*Lampadena luminosa*), Clorophiidae (*Lophius* sp.) and an unidentified Gadidae (Fernández et al., 2009). Based on an observation by Baird et al. (2021) of a dwarf sperm whale sighted off Hawaii, which exhibited scarring on the snout suggestive of bottom-feeding, the extensive and old scarring on the snouts of stranded pygmy sperm whales in the Canary Islands could similarly indicate benthic foraging in these animals.

In accordance with Baird et al. (2021), the evidence suggests that both species are characterized more by their elusiveness than their rarity. Both dwarf and pygmy sperm whales are underestimated in conventional visual surveys, and acoustic detections highlight that the lack of visual sightings does not necessarily mean this genus is not present in a specific area (Baumgartner et al., 2001; Hodge et al., 2018; Hildebrand et al., 2019). Previous acoustic recordings in the south of Fuerteventura Island in deep water (>1000 m) attributed to harbor porpoise (*Phocoena phocoena*) probably could be of *Kogia* (Boisseau et al., 2007). Clicks of both species have similar acoustic characteristics with frequencies between 60–130 kHz for both (Marten, 2000; Madsen et al., 2005; Merkens et al., 2018), but *P. phocoena* has a shorter inter-click interval (ICI) of 40–60 ms compared to 64–84 ms for *Kogia* whales. The stranding record and sightings indicate that, at least during some months, pygmy and dwarf sperm whales share the same waters in the archipelago. Sightings of *Kogia* spp. in the Canary Archipelago were predominantly outside the slope of the islands. The average group size of *K. breviceps* and *K. sima* observed in the Canary Islands is similar to the mean group sizes reported for these species in the literature (Nagorsen, 1985; Best, 2007; Baird, 2016; McAlpine, 2018; Baird et al., 2021; Plön, 2022; Plön and Baird, 2022). The sighting of two adult-sized individuals identified as dwarf and pygmy sperm whales respectively (Fig. 12), is, to our knowledge, the first documented record of an encounter with spatial association of both species.

4.4 Predation

Stranded animals of both species in the Canary Islands exhibited signs of interspecific aggressive interactions or predation by killer whales (*Orcinus orca*), such as cutaneous tooth rake marks, hematomas, external and internal haemorrhages, and bone fractures. Puig-Lozano et al. (2020), in a retrospective study of traumatic intra- and interspecific interactions in a sample of 540 stranded and necropsied cetaceans in the Canary Islands between 2000 and 2017, identified traumatic etiologies such as acute tooth rake marks, skin erosion/laceration, skin vascular changes, fractures mainly affecting the thoracic region, tracheal oedema, and CNS vascular changes, among others, and recorded three cases of predation by killer whales (representing 12.5% of positive cases). During the reevaluation of the available material and photographic verification, we identified nine further cases, in addition to the three described by Puig-Lozano et al. (2020) (ten pygmy sperm whales and two dwarf sperm whales), of probable predation by killer whales, based on the presence of external, acute, punctate tooth rake marks consistent with this predator (Table 3). Eight of these individuals were female, five of which were pregnant. The percentage of simultaneously pregnant and lactating females in both *Kogia* species (Plön, 2004) suggests a potential loss of calves, which may occur during violent interspecific interactions. Whether pregnant or accompanied by calves, these females may represent a critical target for orca attacks or harassment. These interactions have been documented at sea in the waters of the Canary Archipelago, providing evidence of predation pressure from killer whales in the region. A video recording by Canarian artisanal fishermen shows killer whales attacking an unidentified *Kogia* species. In one sequence, an adult male killer whale pushes and grabs the *Kogia* specimen, which releases a cloud of fecal fluid before both animals disappear from the surface. Some carcasses with signs of killer whale predation reach the shore relatively intact, without a significant loss of tissue mass (Figs. 6.5 and 7.1). Along with evidence of stranded animals that survive after their injuries, this suggests that some animals can escape or that the carcasses are discarded and not consumed.

There is a potential for these interactions to go unnoticed or be underestimated, especially in carcasses showing advanced autolysis, postmortem scavenging by sharks resulting in substantial tissue loss, and those displaying trauma without bite tooth marks, such as those caused by headbutts and tail slaps from killer whales. Predation on dwarf sperm whales by killer whales has been documented twice in the Bahamas (Dunphy-Daly et al., 2008). No healed or fresh cookiecutter shark (*I. brasiliensis*) or largetooth cookiecutter shark (*I. plutodus*) bites were observed in the stranded specimens or at sea, which contrasts with observations from other regions such as Hawaii (Baird et al., 2021).

Predation pressures can determine various aspects of the ecology and behavior of kogiid species, which are vulnerable to predators such as sharks and killer whales due to their relatively small size. Plön (2022) hypothesized that bot the presence of false gills and the ability to release ink in kogiids evolved at least in part due to predation pressure, particularly from sharks and killer whales. Baird et al. (2021) described a vigilance behavior in dwarf sperm whales in Hawaii when they are at the surface. The separation between mother and offspring observed in two sightings of dwarf sperm whale off the Canary Islands (Fig. 13) may constitute a potential form of predator mimicry or vigilance behavior in response to predation risk, a mechanism to minimize detection by visually oriented predators such as sharks.

4.5 Conservation

Both species are listed as Least Concern by the International Union for Conservation of Nature (IUCN) Red List (Kiszka and Braulik, 2020). The species are included in the Spanish List of Wild Species under Special Protection Regime but not in the Spanish Catalogue of Endangered Species (Real Decreto 139/2011). Among the threats to the conservation of these species are climate change, acoustic disturbances, fisheries interactions, maritime traffic, and vessel strikes, chemical contaminants, and marine debris pollution (Plön and Relton, 2016). Both species are exposed to actual and potential risks and cumulative effects in the Canary Islands due to anthropogenic activities, including heavy maritime traffic, military sonar, vessel noise, seismic air guns, and offshore wind energy farms. The sea surface temperature in the Canary Current Large Marine Ecosystem (CCLME) over the 32 years from 1982 to 2013 reveals an average warming trend of 0.28 °C per decade. Additionally, in the deep waters of the oceanic areas north of the Canary Islands, the temperature is decreasing at a rate of −0.01 °C per decade and a salinity reduction of −0.002 per decade have been observed (Vélez-Belchí et al., 2015). Although pygmy and dwarf sperm whales, due to their surface temperature preferences, belong to the group of species whose distribution could expand due to the intrusion of warmer waters at more northern latitudes, indirect effects of climate change, such as changes in the distribution and density of prey at different trophic levels of marine ecosystems or the appearance of emerging diseases are less well known (MacLeod, 2009; Plön and Relton, 2016).

Two stranded pygmy sperm whales in the Canary Islands between 2000 and 2015 had foreign bodies (debris) in their stomachs (Puig-Lozano et al., 2018). A 240 cm female live stranded on May 21, 2001, in Gran Canaria (Kbr23), was kept in a tank for 11 days before being returned to the sea. An endoscopic examination revealed a small piece of netting and a cluster of fishing lines in its stomach, which were removed. In a pygmy sperm whale stranded in the Canary Islands, nearly a hundred microplastic fragments were found in its digestive system (Montoto-Martínez et al., 2021). In a retrospective study on fishery interactions in stranded cetaceans in the Canary Islands between January 2000 and December 2018 (Puig-Lozano et al., 2020), no cases involving *Kogia* were found.

Anthropogenic noise produced by high-intensity naval mid-frequency active sonar or seismic surveys can threaten *Kogia* whales. Strandings coincident with naval exercises suggest that pygmy and dwarf sperm whales may be susceptible to impacts from antisubmarine sonar. They have been involved in stranding events associated with naval mid-frequency active antisubmarine sonar use in the Canary Islands (Simmonds and Lopez-Jurado, 1991), Hawaii (Baird, 2016), North Carolina (Hohn et al., 2006), Taiwan (Wang and Yang, 2006; Yang et al., 2008), and Florida (Waring et al., 2006). A live stranding of a 270 cm pregnant female with a 175 cm calf occurred on November 25, 1988, in northeast Lanzarote Island, along with the stranding of three Cuvier's beaked whales (*Z. cavirostris*) and one northern bottlenose whale (*Hyperoodon ampullatus*) on the southeast coast of Fuerteventura Island, coinciding with the naval exercises "FLOTA 88." These events have been associated with high-intensity naval mid-frequency active sonar use (Cox et al., 2006; Hohn et al., 2006; Simmonds and Lopez-Jurado, 1991). In the eastern waters of Lanzarote and Fuerteventura, seismic surveys have been conducted for hydrocarbon and gas exploration, which could potentially impact *Kogia* spp. and beaked whales. The waters off the eastern coast of the islands of Lanzarote and Fuerteventura were designated as a Site of Community Interest (SCI): "Marine Area of Eastern and Southern Lanzarote-Fuerteventura (LIC-ESZZ15002)", as part of the Macaronesian Natura 2000 Network.

Vessel collisions with pygmy sperm whales have been documented (Sylvestre, 1988). Collisions between vessels and whales are a growing conservation issue of concern in the Canary Archipelago, primarily due to the trend of increasing numbers and speeds of ships in recent years. For example, 60% of sperm whale deaths in the Canary Islands are due to ship strikes (Arregui et al., 2019). Carrillo and Ritter (2010) found evidence of ship strikes in 10 pygmy sperm whales (17%) out of 59 that presented deep cuts, partially or entirely consistent with collisions with boats, from a sample of 556 cetacean carcasses found ashore in the Canary Archipelago between

1991 and 2007. However, some of the cases were not diagnosed using standardized forensic techniques. Four vessel collisions involving pygmy sperm whales were reported in the Canary Islands between 2006 and 2012 (Arregui et al., 2019), and fat embolism detection was used to determine if the collisions were antemortem, especially in decomposed carcasses. Due to their prolonged periods at the surface and logging behavior, these animals may be more susceptible to being struck by vessels. Additionally, the small size of *Kogia* species and the open waters of the archipelago suggest that this mortality could be underestimated due to carcass drifting or sinking, raising particular concern. Offshore wind farms are an expanding energy production sector. Their development and deployment on a commercial scale, including in some marine sectors of the Canary Archipelago, are on the rise. However, their potential environmental impacts, particularly on marine ecosystems and cetaceans, remain largely speculative (Galparsoro et al., 2022). Some areas where this industry has been proposed for the future include the oceanic waters off the eastern coast of the islands of Lanzarote and Fuerteventura, an area with many sightings of *Kogia* spp.

5. Conclusions

The combined data from strandings and sightings at sea of kogiids seems to indicate that the Canary Archipelago is an important habitat for pygmy and dwarf sperm whales. In addition, the high stranding rate, particularly of pygmy sperm whales, suggests that the Canary Archipelago is a global stranding hotspot for kogiids.

While *K. breviceps* exhibits a consistent year-round presence in the waters of the Canary Islands, the dwarf sperm whale appears to be present only during certain months of the year.

The strandings of pregnant and lactating females, mother-calf pairs, and neonates, corroborated by observations at sea, confirm that both species reproduce in the waters off the Canary Islands.

Based on the available information, we estimate that the conception period for pygmy sperm whales in the Canary Archipelago occurs from June to February, peaking between September and December. The birth period extends from May to January, with a peak between August and November. Insufficient data are available for the dwarf sperm whale.

Groups of kogiids with calves were found at significantly shallower depths and closer to the coast compared to groups without calves.

The data seem to support the idea that the Canary Archipelago, particularly the eastern waters of Lanzarote and Fuerteventura, is an important feeding area for *Kogia* whales, probably due to a confluence of oceanographic processes.

This study highlights killer whale predation on pygmy and dwarf sperm whales in the Canary Islands, primarily affecting adult females (including a notable number of pregnant individuals). This suggests targeted predation on this sex and age class and potentially underestimated predation pressure on both species.

Both species in the region are subjected to current and potential threats and cumulative impacts from multiple anthropogenic activities, encompassing vessel strikes, heavy maritime traffic, military sonar, vessel-generated noise, seismic air guns, and offshore wind energy installations.

However, information gaps limit our ability to accurately assess population trends in the area for both species, emphasizing the need for further systematic field research.

6. Future perspectives

Monitoring oceanic cetaceans at sea can be costly and challenging, especially for *Kogia*. Their detection is strongly associated with very calm days, encounters are brief, and the animals dive quickly, making them difficult to approach or tag, complicating their study at sea. The offshore waters east of Lanzarote and Fuerteventura present an opportunity for at-sea studies of kogiids, as this area is strategically positioned for field research.

This investigation provides a scientific baseline for future research, which could benefit from the merged use of unmanned aerial systems (UAS drones; Baird et al., 2021) and passive acoustic monitoring methods (bottom and drifting recorders) employed in other regions, such as the western North Atlantic shelf break (Hodge et al., 2018), Gulf of Mexico (Hildebrand et al., 2019), Mariana Archipelago (McCullough et al., 2021), and the Bahamas (Malinka et al., 2021). Additionally, using environmental DNA from surface water samples offers promising research opportunities with both cryptic species, as this technique has already been successfully tested with *K. sima* at Malpelo Island (Eastern Pacific, Colombia; Juhel et al., 2021). Long-term photo identification could help determine residency, population sizes, and trends. Integrating these parameters with biological information from strandings would enhance understanding of both species' life history and population health. The latter includes reproduction, growth, feeding ecology through diet composition from stomach contents, and stable isotope analysis. Furthermore, it is crucial to properly assess anthropogenic pressures and the conservation status of these species in the region, particularly in the marine protected waters off the eastern coast of the islands of Lanzarote and Fuerteventura.

Acknowledgments

This paper is dedicated to the memory of Manuel Carrillo Pérez, Rogelio Herrera, and Bernd Brederlau for their invaluable work in the study and conservation of cetaceans and the marine environment of the Canary Islands. We are grateful to Pascual Calabuig, Cristina Lorenzo, Leire Ruiz, Silvana Neves, Alexis Rivera, Sonia García, Saioa Talavera, Bernd Brederlau, Silvia Hildebrandt, Arquímedes Bermúdez, Marco Mattone and Barbara Gazzetta of Wildrider, Gonzalo Apesteguía of wewhale, Marta Lorente, Teodoro Lucas, Erika Urquiola, Rogelio Herrera, Leopoldo Moro, Manuel Arechavaleta, Ana Carrasco and Francisco Rodríguez for their work and assistance. We are also grateful to all the volunteers who helped us both in the postmortem studies of stranded animals and during fieldwork at sea. We appreciate the contributions of the different members of the Canary Islands Stranding Network, as well as the assistance and support provided by the various island administrations (municipalities and island councils) of the Canary Archipelago, especially to the Biosphere Reserve Office of the Lanzarote Island Council. Our gratitude to Marina Sequeira (Instituto da Conservação da Natureza e das Florestas, Lisbon, Portugal) and Alfredo López (Coordinadora para o Estudo dos Mamíferos Mariños CEMMA, Galicia, Spain) for kindly providing the information on strandings in Portugal and Galicia. The sea fieldwork conducted by SECAC and CEAMAR was funded by various projects from the Canary Government (Consejería de Transición Ecológica y Energía), the Spanish Government (Fundación Biodiversidad and Ministerio para la Transición Ecológica y el Reto Demográfico), and the European Union: Macetus (Feder/Interreg III-B MAC/4.2/M10), Emecetus (Feder/Interreg III-B 05/MAC/4.2/M10), Life (Life03Nat0062), Indemares Life+ (Life07/Nat/E/00732), Mistic Seas II, and Marcet II (Programa Interreg MAC 2014–2020, MAC2/4.6c/392). The data analysis for this work was conducted within the framework of the BIOCETCAN project funded by Consejería de Transición Ecológica y Energía. Permissions required for managing stranded cetaceans and data collection were issued by Ministerio para la Transición Ecológica y el Reto Demográfico del Gobierno de España (MITECO). We are grateful to the reviewers for their constructive comments that helped to improve this manuscript.

References

Albouy, C., Delattre, V., Donati, G., Frölicher, T.L., Albouy-Boyer, S., Rufino, M., et al., 2020. Global vulnerability of marine mammals to global warming. Sci. Rep. 10, 1–12.

Alves, F., Alessandrini, A., Servidio, A., Mendonça, A.S., Hartman, K.L., Prieto, R., et al., 2019. Complex biogeographical patterns support an ecological connectivity network of a large marine predator in the northeast Atlantic. Divers. Distrib. 25 (2), 269–284. https://doi.org/10.1111/ddi.12848.

Arbelo, Espinosa de los Monteros, M., Herráez, A., Andrada, P., Sierra, M., Rodríguez, E., et al., 2013. Pathology and causes of death of stranded cetaceans in the Canary Islands (1999–2005). Dis. Aquat. Organ. 103 (2), 87–99.

Aristegui, J., Sangra, P., Hernández-León, S., Cantón, M., Hernández- Guerra, A., Kerling, J.L., 1994. Island-induced eddies in the Canary Islands. Deep. Sea Res. 41, 1509–1525. https://doi.org/10.1016/0967-0637(94)90058-2.

Ariza, A., Landeira, J.M., Escánez, A., Wienrroither, R., Aguilar De Soto, N., Rostad, A., et al., 2016. Vertical distribution, composition and migratory patterns of acoustic scattering layers in the Canary Islands. J. Mar. Syst. 157, 82–91. https://doi.org/10.1016/j.marsys.

Arregui, M., Bernaldo de Quirós, Y., Saavedra, P., Sierra, E., Suárez-Santana, C.M., Arbelo, M., et al., 2019. Fat embolism and sperm whale ship strikes. Front. Mar. Sci. 6, 379. https://doi.org/10.3389/fmars.2019.00379.

Badcock, J., 1970. The vertical distribution of mesopelagic fishes collected of the Sond Cruise. J. Mar. Ass. U. K. 50, 1001–1044.

Baird, R.W., 2016. The Lives of Hawai]i's Dolphins and Whales: Natural History and Conservation. University of Hawai]i Press.

Baird, R.W., Mahaffy, S.D., Lerma, J.K., 2021. Site fidelity, spatial use, and behavior of dwarf sperm whales in Hawaiian waters: using small-boat surveys, photo-identification, and unmanned aerial systems to study a difficult-to-study species. Mar. Mammal. Sci. 38 (1), 326–348. https://doi.org/10.1111/mms.12861.

Barlow, J., 2015. Inferring trackline detection probabilities, g (0), for cetaceans from apparent densities in different survey conditions. Mar. Mammal. Sci. 31 (3), 923–943. https://doi.org/10.1111/mms.12205.

Barton, E.D., Arístegui, J., Tett, P., Cantón, M., García-Braun, J., Hernández-León, S., et al., 1998. The transition zone of the Canary Current upwelling region. Prog. Oceanogr. 41 (4), 455–504.

Baumgartner, M.F., Mullin, K.D., Nelson May, L., Leming, T.D., 2001. Cetacean habitats in the northern Gulf of Mexico. Fish. Bull. 99, 219–239.

Berini, C.R., Kracker, L.M., McFee, W.E., 2015. Modeling pygmy sperm whale (*Kogia breviceps*, De Blainville 1838) strandings along the southeast coast of the United States from 1992 to 2006 in relation to environmental factors. NOAA Technical Memorandum NOS. NCCOS 203. Charleston, SC. 44. https://doi.org/10.7289/V5/TM-NOS-NCCOS-203.

Best, P.B., 2007. Whales and Dolphins of the Southern African Subregion. Cambridge University Press 338 pages.

Boisseau, O., Matthews, J., Gillespie, D., Lacey, C., Moscrop, A., El Ouamari, N., 2007. A visual and acoustic survey for habour porpoises off North-West Africa: further evidence of a discrete population. Afr. J. Mar. Sci. 29 (3), 1–8. http://dx.doi.org/10.2989/AJMS.2007.29.3.8.338.

Bordes, F., Uiblein, F., Castillo, R., Barrera, A., Castro, J.J., Coca, J., et al., 1999. Epi-and mesopelagic fishes, acoustic data, and SST images collected off Lanzarote, Fuerteventura and Gran Canaria, Canary Islands, during cruise "La Bocaina 0497". Inf. Tec. Inst. Canario Cienc. Mar. 5, 1–45.

Bordes, F., Wienrroither, R., Uiblein, F., Moreno, T., Bordes, C., Hernández, V., et al., 2009. Catálogo de especies meso y batipelágicas. Peces, moluscos y crustáceos. Colectadas con arrastre en las Islas Canarias, durante las campañas realizadas a bordo del B/E "La Bocaina". Instituto Canario de Ciencias Marinas (ICCM), Agencia Canaria de Investigación, Innovación y Sociedad de la información—Gobierno de Canarias, Consejería de Agricultura, Ganadería, Pesca y Alimentación, Viceconsejería de pesca, 326 pp.

Braun, C.D., Arostegui, M.C., Thorrold, S.R., Papastamatiou, Y.P., Gaube, P., Fontes, J., et al., 2022. The functional and ecological significance of deep diving by large marine predators. Ann. Rev. Mar. Sci. 14, 129–159. https://doi.org/10.1146/annurev-marine-032521-103517.

Breese, D., Tershey, B.R., 1993. Relative abundance of Cetacea in the Canal de Ballenas, Gulf of California. Mar. Mammal. Sci. 9 (3), 319–324. https://doi.org/10.1111/j.1748-7692.1993.tb00460.x.

Caldwell, D.K., Caldwell, M.C., 1989. Pygmy sperm whale *Kogia breviceps* (de Blainville, 1838): dwarf sperm whale *Kogia simus* Owen, 1966. In: Ridgway, S.H., Harrison, R. (Eds.), Handbook of Marine Mammals Vol. 4. Academic Press, pp. 235–260.

Carrillo, M., Ritter, F., 2010. Increasing numbers of ships strikes in the Canary Islands: proposals for immediate action to reduce risk of vessel-whale collisions. J. Cetacean Res. Manage. 11 (2), 131–138.

Casinos, A., 1977. On a stranding of a pygmy sperm whale, *Kogia breviceps* (de Blainville, 1838) on the Canary Islands. Saugetierkundliche Mitteilungen 25 (1), 79–80.

Chivers, S.J., LeDuc, R.G., Robertson, K.M., 2005. Genetic variation of *Kogia* spp. With preliminary evidence for two species of *Kogia sima*. Mar. Mammal. Sci. 21 (4), 619–634. https://doi.org/10.1111/j.1748-7692.2005.tb01255.x.

Coombs, E.J., Deadville, R., Sabin, R.C., Allan, L., O´Connell, M., Berrow, S., et al., 2019. What can cetacean stranding record tell us? A study of UK and Irish cetacean diversity over the past 100 years. Mar. Mammal. Sci. 35 (4), 1527–1555. https://doi.org/10.1111/mm.12610.

Cox, T.M., Ragen, T.J., Read, A.J., Vos, E., Baird, R.W., Balcomb, K., et al., 2006. Understanding the impacts of anthropogenic sound on beaked whales. J. Cetacean Res. Manage. 7 (3), 177–187.

De Oliveira Ferreira, N., Santos-Neto, E.B., Manhaes, B.M.R., Carvaho, V.L., GonÇalves, L., De Castillo, P.V., et al., 2023. The deep dive of organohalogen compounds: bioaccumulation in the top predators of mesopelagic trophic webs, pygmy and dwarf sperm whales, from the Southwestern Atlantic Ocean. Chemosphere 345, 140456. https://doi.org/10.1016/J.Chemosphere.2023.140456.

Díaz-Delgado, J., Fernández, A., Sierra, E., Sacchini, S., Andrada, M., Vela, A.I., et al., 2018. Pathologic findings and causes of death of stranded cetaceans in the Canary Islands (2006–2012). PLoS One 13 (10), e0204444. https://doi.org/10.1371/journal.pone.0204444.

Dinis, A., Molina, C., Tobeña, M., Sambolino, A., Hartman, K., Fernández, M., et al., 2021. Large-scale movements of common bottlenose dolphins in the Atlantic: dolphins with an international courtyard. PeerJ 9, e11069. https://doi.org/10.7717/peerj.11069.

Doty, M.S., Oguri, M., 1956. The island mass effect. ICES J. Mar. Sci. 22, 33–37. https://doi.org/10.1093/icesjms/22.1.33.

Dunphy-Daly, M.M., Heithaus, M.R., Claridge, D.E., 2008. Temporal variation in dwarf sperm whale (*Kogia sima*) habitat use and group size off Great Abaco Island, Bahamas. Mar. Mammal. Sci. 24 (1), 171–182. https://doi.org/10.1111/j.1748-7692.2007.00183.x.

Erwin, P.M., Rhodes, R.G., Kiser, K.B., Keenan-Bateman, T.F., McLellan, W.A., Pabst, D.A., 2017. High diversity and unique composition of gut microbiomes in pygmy (*Kogia breviceps*) and dwarf (*K. sima*) sperm whales. Sci. Rep. 7, 7205. https://doi.org/10.1038/s41598-017-07425-z.

Escánez, A., Guerra, A., Riera, R., Ariza, A., González, A.F., Aguilar de Soto, N., 2022. New contribution to the knowledge of the mesopelagic cephalopod community off the western Canary Islands slope. Reg. Stud. Mar. Sci. 55, 102572. https://doi.org/10.1016/J.rsma 2022.102572.

Fernández, R., Santos, M.B., Carrillo, M., Tejedor, M., Pierce, G., 2009. Stomach contents of cetaceans stranded in the Canary Islands 1996–2006. J. Mar. Biol. Assoc. U. K. 89 (5), 873–883.

Ferreira, R., Steiner, L., Martín, V., Fusar Poli, F., Dinis, A., Kaufmann, M., et al., 2022. Unraveling site fidelity and residency patterns of sperm whales in the insular oceanic waters of Macaronesia. Front. Mar. Sci. 9, 1021635. https://doi.org/10.3389/fmars.2022.1021635.

Galparsoro, I., Menchaca, I., Garmendia, J.M., Borja, A., Maldonado, A.D., Iglesias, G., et al., 2022. Reviewing the ecological impacts of offshore wind farms. npj Ocean. Sustain. 1, 1. https://doi.org/10.1038/s44183-022-00003-5.

Geraci, J.R., Lounsbury, V.J., 2005. Marine Mammals Ashore: A Field Guide for Strandings, second ed. National Aquarium in Baltimore, Baltimore, MD.

Gómez-Lobo, D.A., Monteoliva, A.P., Fernández, A., Arbelo, M., de la Fuente, J., Pérez-Gil, M., et al., 2024. Mitochondrial Variation of Bottlenose Dolphins (*Tursiops truncatus*) from the Canary Islands Suggests a Key Population for Conservation with High Connectivity within the North-East Atlantic Ocean. Animals 14 (6), 901. https://doi.org/10.3390/ani14060901.

Handley, C.O., 1966. A synopsis of the genus *Kogia* (pygmy sperm whales). In: Norris, K.S. (Ed.), Whales, Dolphins and Porpoises. University of California Press, pp. 62–69.

Hernández-García, V., Martín, V., 1996. Food habits of the pygmy sperm whale Kogia breviceps (de Blainville, 1838) based on stranded animals in the Canary Islands. II Symposium "Fauna and Flora of the Atlantic Islands". 12–16 de febrero de 1996. Las Palmas de Gran Canaria.

Hernández-Guerra, A., Fraile-Nuez, E., Borges, R., López-Laatzen, F., Vélez-Belchí, P., Parrilla, G., et al., 2003. Transport variability in the Lanzarote passage (eastern boundary current of the North Atlantic subtropical Gyre). Deep-Sea Res. Part. I: Oceanographic Res. Pap. 50 (2), 189–200. https://doi.org/10.1016/S0967-0637(02)00163-2.

Heyning, J.E., 1997. Sperm whale phylogeny revisited: analysis of the morphological evidence. Mar. Mammal. Sci. 13 (4), 596–613.

Hildebrand, J.A., Frasier, K.E., Baumann-Pickering, S., Wiggins, S.M., Merkens, K.P., Garrison, L.P., et al., 2019. Assessing seasonality and density from passive acoustic monitoring of signals presumed to be from pygmy and dwarf sperm whales in the Gulf of Mexico. Front. Mar. Sci. 6, 66. https://doi.org/10.3389/fmars.2019.00066.

Hodge, L.E.W., Baumann-Pickering, S., Hildebrand, J.A., Bell, J.T., Cummings, E.W., et al., 2018. Heard but not seen: occurrence of *Kogia* spp. along the western North Atlantic shelf break. Mar. Mamm. Sci. 34, 1141–1153. https://doi.org/10.1111/mms.12498.

Hohn, A.A., Rotstein, D.S., Harms, C.A., Southall, B.L., 2006. Report on marine mammal unusual mortality event UMESE0501Sp.: multispecies mass stranding of pilot whales (*Globicephalamacrorhynchus*), minke whale (*Balaenoptera acutorostrata*), and dwarf sperm whales (Kogia sima) in North Carolina on 15–16. pp. 1–222. NMFS-SEFSC-537. NOAA Technical Memorandum.

Hutterer, R., 1994. Dwarf sperm whale *Kogia simus* in the Canary Islands. Lutra 37, 89–92.

Jefferson, T.A., Webber, M.A., Pitman, R.L., 2015. Marine Mammals of the World: A Comprehensive Guide to Their Identification, second ed. Academic Press.

Joblon, M.J., Pokras, M.A., Morse, B., Harry, C.T., Rose, K.S., Sharp, S.M., et al., 2014. Body condition Scoring system for Delphinids based on short-beaked common dolphins (*Delphinus delphis*). J. Mar. Anim. Their Ecol. 7 5e13.

Juhel, J.B., Marques, V., Fernández, A.P., Borrero-Pérez, G.H., Martinguerra, M.M., Valentini, A., et al., 2021. Detection of elusive dwarf sperm whale (*Kogia sima*) using environmental DNA at Mapelo island (Eastern Pacific, Colombia). Ecol. Evol. Nat. Note. Nat. notes. 11, 2956–2962. https://doi.org/10.1002/ECE3.7057.

Kasuya, T., 1977. Age determination and growth of the Baird's beaked whale, with a comment on the fetal growth rate. Sci. Rep. Whales Res. Inst. 29, 1–20.

Keenan-Bateman, T.F., McLellan, W.A., Harms, C.A., Piscitelli, M.A., Barco, S.G., Thayer, V.G., et al., 2016. Prevalence and anatomic site of *Crassicauda* sp. infection, and its use in species identification, in kogiid whales from the mid-Atlantic United States. Mar. Mam. Sci. 32, 868–883. https://doi.org/10.1111/mms.12300.

Kiszka, J., Braulik, G., 2020. *Kogia sima*. IUCN Red List of Threatened Species 2020, e. T11048A50359330. https://doi.org/10.2305/IUCN.UK.2020.2.RLTS.T11048A50359330.

Kuiken, T., García-Hartmann, M., 1991. Proc 1st ECS Workshop on Cetacean Pathology: Dissection Techniques and Tissue Sampling, vol. 17. European Cetacean Society Newsletter, Saskatoon. Spec Issue. pp. 1–39.

MacLeod, C.D., 2009. Global climate change, range changes and potential implications for the conservation of marine cetaceans: a review and synthesis. Endang. Species Res. 7, 125–136. https://doi.org/10.3354/esr00197.

Madsen, P.T., Carder, D.A., Bedholm, K., Ridgway, S.H., 2005. Porpoise clicks from a sperm whale nose—convergent evolution of 130 kHz pulses in toothed whales? Bioacoustics 15, 195–206.

Malinka, C.E., TØnnensen, P., Dunn, C.A., Claridge, D.E., Gridley, T., Ellen, S.H., et al., 2021. Echolocation click parameters and biosonar behaviour of the dwarf sperm whale (*Kogia sima*). J. Exp. biol. 24, JEB 240689. https://doi.org/10.1242/jeb.240689.

Mann, K., Lazier, J., 1991. Dynamics of Marine Ecosystems: Biological–Physical Interactions in the Oceans. Blackwell, London.

Marten, K., 2000. Ultrasonic analysis of pygmy sperm whale (*Kogia breviceps*) and Hubbs' beaked whale (*Mesoplodon carlhubbsi*) clicks. Aquat. Mamm. 26 (1), 45–48.

McAlpine, D.F., 2014. Family Kogiidae (pygmy Sperm Whales). In: Wilson, D.E., Mittermeier, R.A. (Eds.), Handbook of the Mammals of the World Vol. 4. Lynx Edicions, Barcelona, pp. 318–325.

McAlpine, D.F., 2018. Pygmy and dwarf sperm whales *Kogia breviceps* and *Kogia sima*. In: Würsig, B., Thewissen, J.G.M., Kovacs, K.M. (Eds.), Encyclopedia of Marine Mammals, third ed. Academic Press/Elsevier, pp. 786–788.

McAlpine, D.F., Murison, L.D., Hoberg, E.P., 1997. New records of the pygmy sperm whale, *Kogia breviceps* (Physeteridae) from Atlantic Canada with notes on diet and parasites. Mar. Mamm. Sci. 13 (4), 70–704.

McCullough, J.L.K., Wren, J.L.K., Oleson, E.M., Allen, A.N., Siders, Z.A., Norris, E.S., 2021. An acoustic survey of beaked whales and *Kogia* spp. in the Mariana Archipelago using drifting recorders. Front. Mar. Sci. 8, 664292. https://doi.org/10.3389/FMARS.20.8664292.

McLuor, A.J., Williams, C.T., Alves, F., Dinis, A., Pais, M.P., Canning-Clode, J., 2022. The status of marine megafauna research in macaronesia: A systematic review. Front. Mar. Sci. 9, 819581. https://doi.org/10.3389/fmars.2022.819581.

Merkens, K., Mann, D., Janik, V.M., Claridge, D., Hill, M., Oleson, E., 2018. Clicks of dwarf sperm whales (*Kogia sima*). Mar. Mammal. Sci. 34 (5), 963–978. https://doi.org/10.1111/mms.12488.

Montoto-Martínez, T., de La Fuente, J., Puij-Lozano, R., Marques, N., Arbelo, M., Hernández-Brito, J.J., et al., 2021. Microplastics, bisphenols, phalates and pesticides in odontocete species in the Macaronesian region (Eastern North Atlantic). Mar. Pollut. Bull. 173, 113105.

Moura, J.F., Acevedo-Trejos, E., Tavares, D.C., Meirelles, A.C.O., Silva, C.P.N., Oliveira, L.R., et al., 2016. Stranding events of *Kogia* whales along the Brazilian coast. PLoS One 11 (1), e0146108. https://doi.org/10.1371/journal.pone.0146108.

Nagorsen, D., 1985. Kogia simus. Mamm. Species. 239, 1–6.

Norris, K.S., 1961. Standardized methods for measuring and recording data on the smaller cetaceans. J. Mammal. 42 (4), 471–476.

Norris, K., Terry, A., Hansford, J.P., Turvey, S.T., 2020. Biodiversity conservation and the earth system: mind the gap. Trends Ecol. Evol. 35, 919–926. https://doi.org/10.1016/j.tree.2020.06.010.

Odell, D.K., 1991. A review of the Southeastern United States Marine Mammal Stranding Network: 1978–1987. In: Marine Mammal Strandings in the United States. NOAA Technical Report NMFS 98. Proceedings of the Second Marine Mammal Stranding Workshop, Miami, FL, pp. 19–23.

Ogawa, T., 1936. Studien über die zahnwale in Japan. Pt. IV Cogia (In Japanese). Botany Zool. 4, 2017–2024.
Plön, S., 2004. The Status and Natural History of Pygmy (*Kogia breviceps*) and Dwarf (*K. sima*) Sperm Whales Off Southern Africa (PhD thesis). Rhodes University, Grahamstown, South Africa. 551 pp.
Plön, S., 2022. Pygmy sperm whale *Kogia breviceps* (de Blainville, 1838). In: Hackländer, K., Zachos, F.E. (Eds.), Handbook of the Mammals of Europe. Springer, Cham. https://doi.org/10.1007/978-3-319-65038-890-1.
Plön, S., Baird, R.W., 2022. Dwarf sperm whale *Kogia sima* (Owen, 1866). In: Hackländer, K., Zachos, F.E. (Eds.), Handbook of the Mammals of EuropeSpringer, Cham. https://doi.org/10.1007/978-3-319-65038-891-1.
Plön, S., Best, P.B., Duignan, P., Lavery, S.D., Bernard, R.T.F., Waerebeek, K.V., et al., 2023. Population structure of pygmy (*Kogia breviceps*) and dwarf (*Kogia sima*) sperm whales in the Southern Hemisphere may reflect foraging ecology and dispersal patterns. Adv. Mar. Biol. 96, 85–114. https://doi.org/10.1016/bs.amb.2023.09.001.
Plön, S., Relton, C., 2016. A conservation assessment of *Kogia* spp. In: Child, M.F., Roxburgh, L., Do Linh San, E., Raimondo, D., Davies-Mostert, H.T. (Eds.), The Red List of Mammals of South Africa, Swaziland and LesothoSouth African National Biodiversity Institute and Endangered Wildlife Trust, South Africa. https://www.researchgate.net/publication/311677955.
Puig-Lozano, R., de Quirós, Y.B., Díaz-Delgado, J., García-Álvarez, N., Sierra, E., De la Fuente, J., et al., 2018. Retrospective study of foreign body-associated pathology in stranded cetaceans, Canary Islands (2000–2015). Environ. Pollut. 243, 519–527.
Puig-Lozano, R., Fernández, A., Saavedra, P., Tejedor, M., Sierra, E., De la Fuente, J., et al., 2020. Retrospective study of traumatic intra-interspecific interactions in stranded cetaceans, Canary Islands. Front. Vet. Sci. 7, 107. https://doi.org/10.3389/fvets.2020.00107.
Puig-Lozano, R., Fernández, A., Sierra, E., Saavedra, P., Suárez-Santana, C.M., De la Fuente, J., et al., 2020. Retrospective study of fishery interactions in stranded cetaceans, Canary Islands. Front. Vet. Sci. 7, 567258. https://doi.org/10.3389/fvets.2020.567258.
Real Decreto 139/2011, de 4 de febrero, para el desarrollo del Listado de Especies Silvestres en Régimen de Protección Especial y del Catálogo Español de Especies Amenazadas. 2011. Ministerio de Medio Ambiente, y Medio Rural y Marino Boletín Oficial del Estado, 23 febrero 2011; núm. 46. Disponible en: https://www.boe.es/eli/es/rd/2011/02/04/139.
Rice, D.W., 1998. Marine mammals of the world- systematics and distribution. Special Publication No. 4. The Society of Marine Mammalogy. Allen Press, Lawrence, KS, pp. 231.
Roman, J., Estes, J.A., Morissette, L., Smith, C., Costa, D., McCarthy, J., et al., 2014. Whales as marine ecosystem engineers. Front. Ecol. Environ. 12 (7), 377–385. https://doi.org/10.1890/130220.
Ross, G.J.B., 1979. Records of pygmy and dwarf sperm whales, genus *Kogia*, from southern Africa, with biological notes and some comparisons. Ann. Cape Prov. Mus. (Nat. History) 11, 259–327.
Ross, G.J.B., 1984. The smaller cetaceans of the south-east coast of southern Africa. Ann. Cape Prov. Mus. (Nat. History) 15, 173–410.
Santos, M.B., Martín, V., Arbelo, M., Fernández, A., Pierce, G.J., 2007. Insights into the diet of beaked whales from atypical mass stranding in the Canary Islands in September 2002. J. Mar. Biol. Assoc. U. K. 86 (5438), 1–9.
Scott, M.D., Cordaro, J.G., 1987. Behavioral observations of the dwarf sperm whale *Kogia simus*. Mar. Mammal. Sci. 3 (4), 353–354. https://doi.org/10.1111/j.1748-7692.1987.tb00322.x.
Simmonds, M.P., Lopez-Jurado, L.F., 1991. Whales and the military. Nature 351 (6326), 448.

Staudinger, M.D., McAlarney, R.J., McLellan, W.A., Pabst, D.A., 2014. Foraging ecology and niche overlap in pygmy (*Kogia breviceps*) and dwarf (*Kogia sima*) sperm whales from waters of the U.S. mid-Atlantic coast. Mar. Mammal. Sci. 30 (2), 626–655. https://doi.org/10.1111/mms.12064.

Sylvestre, J.P., 1988. On a specimen of pygmy sperm whale *Kogia breviceps* (Blainville, 1838) from New Caledonia. Aquat. Mammal. 14 (2), 76–77.

Tasch, A.C., Lima, R.C., Secchi, E.R., Botta, S., 2023. Niche partitioning among odontocetes in a marine biogeographical transition zone of the western south Atlantic Ocean. Mar. Biol. 171, 38. https://doi.org/10.1007/s00227-023-04359-1.

Thompson, K.F., Millar, C.D., Baker, C.S., Dalebout, M., Steel, D., van Helden, A.L., et al., 2013. A novel conservation approach provides insights into the management of rare cetaceans. Biol. Cons. 157, 331–340.

Vélez-Belchí, P., González-Carballo, M., Pérez-Hernández, M.D., Hernández-Guerra, A., 2015. Open ocean temperature and salinity trends in the Canary Current Large Marine Ecosystem. In: Valdés, L., Déniz-González, I. (Eds.), Oceanographic and Biological Features in the Canary Current Large Marine Ecosystem 115. pp. 299–308 (IOC-UNESCO, Paris. IOC Technical Series, No).

Wang, J.Y., Yang, S., 2006. Unusual cetacean stranding events of Taiwan in 2004 and 2005. J. Cetacean Res. Manage. 8 (3), 283–292.

Waring, G.T., Josephson, E., Fairfield, C.P., Maze-Foley, K., 2006. U.S. Atlantic and Gulf of Mexico marine mammal stock assessments 2005. NOAA Technical Memorandum NMFS-NE 1–346.

West, K., Walker, W., Baird, R., White, W., Levine, G., Brown, E., et al., 2009. Diet of pygmy sperm whales (*Kogia breviceps*) in the Hawaiian Archipelago. Mar. Mamm. Sci. 25 (4), 931–943. https://doi.org/10.1111/j.1748-7692.2009.00295.x.

Willis, P.M., Baird, R.W., 1998. Status of the dwarf sperm whale *Kogia simus* with special reference to Canada. Can. Field—Naturalist 112, 114–125.

Yamada, M., 1954. Some remarks on the pygmy sperm whale *Kogia*. Sci. Rep. Whales Res. Inst. Tokyo. 9, 37–58.

Yang, W., Chou, L.S., Jepson, P.D., Brownell, R.L., Cowan, D., Chang, P.H., et al., 2008. Unusual cetacean mortality event in Taiwan, possibly linked to naval activities. Vet. Rec. 162 (6), 184–186. https://doi.org/10.1136/vr.162.6.184.

CHAPTER THREE

Records from visual surveys, strandings and eDNA sampling reveal the regular use of Reunion waters by dwarf sperm whales

Violaine Dulau[a,*,1], Vanessa Estrade[a,1], Aymeric Bein[b], Natacha Nikolic[c,d], Adrian Fajeau[a], Jean-Marc Gancille[a], Julie Martin[a], Emmanuelle Leroy[a], and Jean-Sebastien Philippe[b]

[a]GLOBICE-Reunion, Grand-Bois, Saint Pierre, Reunion
[b]BIOTOPE, Saint André, Reunion
[c]Agence de Recherche pour la Biodiversité à la Réunion (ARBRE), Saint Gilles, Reunion
[d]INRAE, ECOBIOP, AQUA, Saint-Pée-sur-Nivelle, France
*Corresponding author. e-mail address: violaine.dulau@globice.org

Contents

1. Introduction	66
2. Material and methods	69
2.1 Boat-based and aerial surveys	69
2.2 Environmental DNA (eDNA)	71
2.3 Strandings	73
3. Results	74
3.1 Distribution surveys	74
3.2 Environmental DNA	77
3.3 Strandings	78
4. Discussion	85
4.1 Species occurrence	86
4.2 Spatial distribution around Reunion	87
4.3 Habitat use	89
4.4 Group size and biology	90
4.5 Threats	92
5. Conclusions	93
Acknowledgments	94
Appendix A. Supporting information	94
References	94

[1] Equal contribution.

Abstract

The genus *Kogia* includes two extant species, the dwarf sperm whale (*Kogia sima*) and the pygmy sperm whales (*K. breviceps*). Due to their elusive behavior at the surface, which limits opportunities for observation, they are amongst the least known species of cetaceans and knowledge of their ecology mostly comes from stranded individuals. Although they have overlapping ranges, dwarf sperm whales seem to be distributed preferentially in warmer tropical and subtropical waters, while pygmy sperm whales tend to be associated with more temperate waters. Both species have previously been recorded in the western Indian Ocean, but little is known about their distribution patterns. Data from different sources, including vessel-based and aerial surveys, environmental DNA and strandings were compiled to report on the occurrence of *Kogia* around the remote oceanic island of Reunion. The combination of sightings data, eDNA detections and stranding events indicated that the dwarf sperm whale was more common than the pygmy sperm whale and seems to use the territorial waters of Reunion on a regular basis. The northern part of the island in particular might provide suitable habitats for the species. Groups of 1–5 individuals were sighted and occurred mainly over the insular slope, in 1310 m deep waters and 8.2 km from the shore on average; no clear seasonality pattern could be determined. Stranding data were consistent with a calving period during the austral summer and highlighted the vulnerability of these species to human activities.

1. Introduction

When first described, the dwarf sperm whale (*Kogia sima*, Owen, 1866) and the pygmy sperm whale (*Kogia breviceps*, Blainville, 1838) were recognized as types of sperm whale and included in the Physeteridae family, together with the sperm whale (*Physeter macrocephalus*). Their inclusion was based on several similar characteristics, such as the presence of a spermaceti organ, teeth in the lower jaw only and the asymmetrical alignment of the left blowhole. The genera *Kogia* and *Physeter* have since been revised into two distinct Kogiidae and Physeteridae families (Rice, 1998). Recent phylogenetic and morphological studies support the close relationship between *Kogia* and sperm whales, based on mitochondrial DNA and the anatomical structure of the head (Clarke, 2003; May-Collado and Agnarsson, 2006), further supporting their grouping within the sperm whale superfamily Physeteroidea.

The genus *Kogia* includes two extant species, the dwarf (*K. sima*) and the pygmy (*K. breviceps*) sperm whales, which were definitively identified as separate species in 1966 (Handley, 1966). The two species have overlapping range and occur in tropical and temperate waters of the Atlantic, Indian and Pacific Oceans (Rice, 1998), although dwarf sperm whales seem to prefer warmer tropical and subtropical waters and pygmy sperm whales are usually associated with higher latitudes (Bloodworth and Odell, 2008; Kiszka and Braulik 2020a,b; McAlpine, 2009; Moura et al., 2016). Recent genetic analyses suggested that dwarf sperm whales may actually consist of two parapatric species occupying the Atlantic and the Indo-Pacific Oceans, with the Cape of Good Hope in South Africa representing a barrier (Chivers et al., 2006; Plön et al., 2023a). However, recognition of a putative third *Kogia* species requires further supporting evidence (McAlpine, 2009).

Both the dwarf and pygmy sperm whales were recently listed as "Least Concern" on the IUCN Red List of Threatened Species based on their wide distribution, indications that they might not be as uncommon as initially thought, and the lack of evidence that they are facing major threats (Kiszka and Braulik 2020a,b). However, due to their elusive surface behavior, *Kogia* represents one of the least known of the nine extant families of toothed whales (Baird et al., 2021). They are typically motionless at the surface (logging behavior) or are slow moving (no splash), produce no visible blow and dive by sinking into the water, without showing their flukes, thus providing few cues for detection and making them relatively inconspicuous at the surface unless in very good sea conditions (Baird et al., 2021; Caldwell and Caldwell, 1989). As there are so few at-sea sightings, they are usually described as rare. However, it is not always clear if this is due to their low detectability or low densities. High numbers of strandings and relatively high sighting frequencies (compared to other cryptic species such as beaked whales) which have been reported around some oceanic islands, along continental slope margins and canyons in association with upwelling (Anderson, 2005; Baird et al., 2021; Bloodworth and Odell, 2008; Collins et al., 2002; Dunphy-Daly et al., 2008; Hodge et al., 2018; Plön 2004) suggest that certain areas might provide suitable habitats for populations to establish.

Very few at-sea studies focusing on *Kogia* have been conducted (Baird, 2005, Baird et al., 2021; Dunphy-Daly et al., 2008) and there are few local population estimates, most of which do not discriminate between the two species (Barlow, 2006; Garrison et al., 2010; Laran et al., 2017; Mullin and Fulling, 2004). Species abundance at a global scale is currently unknown and knowledge of population structure, seasonality, and movement patterns is also very limited. Most information on the biology and ecology of these species comes from the analysis of stranded animals (e.g., Plön, 2004; Ross, 1979). Analysis of stomach contents indicates that *Kogia* feed primarily on cephalopods and suggest that each species might use a different ecological niche, with dwarf sperm whales foraging on smaller squid inshore and in shallower depths than pygmy sperm whales (Ross, 1979; Willis and Baird, 1998). Social segregation might also occur, as analysis of prey indicates that females with their young, and immature individuals, might feed on smaller-sized cephalopods closer inshore, than mature males and non-reproductive mature females (Plön, 2004; Ross, 1984). The social structure of *Kogia* remains largely unknown.

Recently, environmental DNA techniques have been shown to be an effective means to assess the distribution of cryptic species such as *Kogia*, as they allow for detection of species presence without relying on direct observations of the animals at the surface (Hodge et al., 2018; Juhel et al., 2021). Kogia can also be detected using passive acoustic monitoring methods (Hildebrand et al., 2019; McCullough et al., 2021; Ridgway and Carder, 2001), although it is not yet possible to acoustically discriminate between the two species, both of which produce high frequency clicks with peak frequencies around 123–130 kHz (Madsen et al., 2005; Malinka et al., 2021; Marten, 2000; Merkens et al., 2018). Tonal calls have not been reported for this species.

In the south-western Indian Ocean (SWIO), the occurrence of both species of *Kogia* has been reported but little is known about their ecology. Historically, relatively high numbers of strandings of both species have been reported along the south-eastern coast of South Africa (106 *K. breviceps* and 85 *K. sima* between 1880 and 1995, Plön, 2004; Ross, 1979). Vessel-based surveys indicated that dwarf sperm whales appear to be relatively common north of the Seychelles (Ballance and Pitman, 1998) and around the Maldives (representing 4.2% of all on-effort sightings), while *K. breviceps* was not recorded (Anderson, 2005). Elsewhere in the region, sightings and strandings of both species of *Kogia* have been reported in lower numbers (Cerchio et al., 2022; Dulau-Drouot et al., 2008; Kiszka et al., 2010; Plön et al., 2023b). Large scale aerial surveys conducted in the SWIO have produced population estimates for both species combined (corrected for availability bias), and showed that although *Kogia* were among the least abundant cetacean taxa, higher densities were observed around the Seychelles (population estimates of 305 individuals), compared to other oceanic islands of the south-western Indian Ocean (Reunion, Mauritius, Mozambique Channel islands) and Madagascar (Laran et al., 2017).

This study aims to investigate the occurrence of *Kogia* around the island of Reunion in the south-western Indian Ocean. Reunion is located at a latitude of ~21°S, at the southern end of the Mascarene Plateau, which extends approximately ~2000 km from the Seychelles to the Mascarene islands (Reunion, Mauritius, Rodrigues). Unlike the Seychelles, which have a continental origin, the Mascarenes are volcanic islands originating from the still-active volcanic hotspot under Reunion. Being geologically young (~3 million years ago), the island of Reunion has a very narrow shelf and steep slope, with depths increasing rapidly to 3000 m. This topography brings oceanic cetacean species in close proximity to the shore, facilitating opportunities for studying deep-sea species that are typically difficult to access. In this study, the occurrence of *Kogia* around Reunion Island was investigated by collating data on the presence of these cryptic species from multiple sources, including visual surveys (both boat-based and aerial), stranding records, and environmental DNA. The objective was to provide new insights into the distribution and phenology of these poorly documented species and highlight potential threats that they may face in these waters.

2. Material and methods
2.1 Boat-based and aerial surveys

Sightings data of *Kogia* were compiled from different type of surveys, including boat-based surveys conducted around the island and aerial surveys conducted off the northern part of the island.

Different types of boat-based cetacean surveys were conducted around Reunion over a 13-year period (Table 1). During 2010–2023, one-day surveys were conducted on small vessels (5–6 m) that covered coastal waters to a distance of ~10 km from shore, without pre-defined transects. Surveys were conducted year-round, at an average of three surveys per week. Surveys were launched from different harbors located mainly on the western side of the island, where boats were more available and weather conditions more favorable. Survey tracks and sighting positions were recorded using a portable GPS unit. Sighting conditions were recorded every 15 min using a 5-level Visibility Index (V): 1: very poor (Beaufort > 4, dawn or dusk); 2: poor (Beaufort 4, numerous white caps and/or poor light), 3: moderate (Beaufort 2 to 3, scattered white-cap, moderate swell), 4: good (Beaufort 2, no or very long swell), 5: excellent (Beaufort 0–1, no swell). Upon sighting a group of cetaceans, the time, GPS position, species identification, group size estimate and behavior data were recorded on a standardized datasheet. Photographs were taken to confirm species identification using a digital SLR camera equipped with a

Table 1 Summary the survey effort per survey type: number of daily surveys, total survey effort (in km) and percentage of effort in good to excellent sighting conditions (visibility index 4 to 5), and number of sightings of dwarf sperm whales (*K. sima*), pygmy sperm whale (*K. breviceps*) and *Kogia sp.* (not identified to species level) reported between 2010 and 2023.

Survey type	Methodology	Survey area	Period	N of daily surveys	Survey effort (km)	% in good to excellent conditions	N of *K. sima* sightings	N of *K. breviceps* sightings	N of *Kogia sp.* sightings
Boat-based	No pre-defined transects	Reunion's coastal waters (up to 10 km)	2010–2023	1915	74,974	78%	4	1	0
Boat-based	Line-transect	Reunion's territorial waters (up to 22 km)	2010–2020	42	4356	80%	1	0	0
Boat-based	Line-transect	Northern part of Reunion (up to 37 km)	2015–2022	379	21,873	75%	19	0	1
Aerial	Line-transect	Northern part of Reunion (up to 50 km)	2015–2022	262	87,588	80%	14	0	44
TOTAL				2840			38	1	45
eDNA		Reunion's territorial waters (up to 22 km)	2018–2022	33			1	1	0

For the eDNA study, effort and *Kogia* detections are reported in numbers of sampling locations (*off survey effort*).

150–300 mm or 150–600 mm lens. Identification to species level was based upon good quality photographs and sightings for which species identification was uncertain were assigned to *Kogia* sp.

Line-transect surveys were also conducted around the island once annually between 2010 and 2020, using a 11 m long fishing vessel with an observation platform located 4 m above the sea surface. Transects covered territorial waters up to 22 km offshore, with designs constrained by the locations of harbors. Each survey was conducted over a 4–5 day period allowing a complete circumnavigation of the island. Due to weather constraints, these 4–5 days surveys around the island could not be completed at the same time every year. Vessel tracks, survey effort and sighting data were recorded using the Logger2000 © interface developed by the International Fund for Animal Welfare (IFAW).

Increased survey effort was completed in the northern part of the island, linked to the construction of a new coastal road viaduct along 12 km of coastline. Both boat-based and aerial surveys were conducted in the area of the construction between January 2015 to June 2022, using pre-defined line-transects. Weather permitting, four boat-based surveys were conducted per month, with transects covering waters up to 37 km offshore, based on the same methodology described above on a 12 m vessel. In addition, two aerial surveys were conducted each month onboard an ultra-light aircraft flying at 300 m altitude that covered waters up to 50 km from shore. As for boat-based survey, the survey track, visibility index (recorded every 15 min using the same five categories as described above), and cetacean sighting positions were collected using a portable GPS and standardised datasheet. Upon sighting, the aircraft left the transect and circled-back to the group to facilitate photographs for species identification and collection of group size estimates, following which the transect was resumed.

Survey effort was reported in a 2 km x 2 km grid, as the cumulative distance (in km) searched in each cell, for boat-based and aerial surveys separately. Spatial distribution was assessed by conducting a kernel density estimation using the sightings data for all surveys combined (boat-based and aerial surveys), using the package adehabitatHR in Rstudio (v2022.07.2). Distribution maps, depth and distance to the nearest coast of each sighting were estimated in QGis (v3.24.3), using bathymetry data provided by the French Naval Hydrographic Services (SHOM).

2.2 Environmental DNA (eDNA)

Seawater samples were collected for eDNA analyses during dedicated surveys from June 2018 to May 2019 and from February 2021 to June 2022 at 34 locations around Reunion, within waters up to 10 km from shore (Fig. 5). Sampling occurred in different months, based on vessel availability and weather conditions and each location was sampled only once. Samples were collected by two trained operators wearing protective clothing and gloves to avoid contamination. At each sampling location, three replicates of 10 liters (L) of surface water were collected and filtered onboard using sterile filter capsules (Sterlitech 0.8 µm,), sterile and single-use tubes and a peristaltic pump. Filter capsules were preserved in RNAlater buffer (Qiagen©) and stored at −20 °C until they could be sent to a dedicated eDNA laboratory (Nature Metrics, UK) for genetic analysis. DNA from each filter was extracted using a DNA extraction kit from Nature Metrics Mammal. Specific primers (MiMammal-UF: 5′-GGG TTG GTA AAT TTC GTG CCA GC-3′, MiMammal-UR: 5′-CAT AGT GGG GTA TCT AAT CCC AGT TTG-3′) were used to amplify ~230 bp of the hypervariable region of the 12S rRNA gene (Ushio et al., 2017).

DNA amplifications were performed with 12 PCR replicates in a final volume of 10 µL (Nikolic et al., submitted). PCR Master Mix (Thermo Scientific) included 0.4 µM of each of the tailed primers, 2 µM of a human blocking primer, 0.8 µg/µL bovine serum albumin (BSA—Thermo Scientific), 3% of DMSO (Thermo Scientific), 1.5 mM of MgCl2 (Invitrogen), and topped up with PCR grade water (Thermo Scientific). PCR conditions were as follows: an initial denaturation at 98 °C for 3 min, followed by 45 cycles of 20 s at 98 °C, 15 s at 69 °C, and 15 s at 72 °C, and a final elongation step at 72 °C for 5 min. Three negative extraction controls and three negative PCR controls (ultrapure water, 12 replicates) were used to assess potential contamination. Amplification success was determined by gel electrophoresis. DNA was purified to remove PCR inhibitors using a DNeasy PowerClean Pro Cleanup Kit (Qiagen). Purified DNA extracts were quantified using a Qubit dsDNA HS Assay Kit on a Qubit 3.0 fluorometer (Thermo Scientific). PCR replicates were pooled and sequencing adapters were added. The final library was sequenced using an Illumina MiSeq V2 kit at 15 pM (Illumina, San Diego, CA, USA) with a 10% PhiX spike.

Metabarcoding pipelines was processed using a NatureMetrics © custom bioinformatics pipeline including quality filtering, trimming, merging paired ends, removal of sequencing errors such as chimeras, clustering of similar sequences into molecular Operational Taxonomic Units (OTUs; each of which approximately represents a species), and matching one sequence from each cluster against the National Center for Biotechnology Information

(NCBI) nucleotide database (GenBank—https://www.ncbi.nlm.nih.gov), using the online nucleotide Basic Local alignment Search Tool (BLAST) and with the default algorithm parameters (https://blast.ncbi.nlm.nih.gov/Blast.cgi). The NCBI database included one referenced sequence for the 12S rRNA region of interest (100% coverage) for *K. sima* (Accession number: NC 041303.1, Shan et al., 2019) and *K. breviceps* (Accession number: AJ554055.1, Arnason et al., 2004), so each *Kogia* species could theoretically be detected. Assignments were made to the lowest possible taxonomic level where there was consistency in the matches and a 100% similarity required for assignment at the species level. Results were presented in terms of total number of reads for each detected species in each sample. For each sampling site, the sample showing the highest number of cetacean reads (out of the three replicates) was reported.

2.3 Strandings

Stranding events and responses in Reunion have been coordinated by GLOBICE since 2006, under the authority of the French stranding network lead by Pelagis laboratory. Post-mortem investigation and tissue sampling were based on the French stranding network's protocols. Data were collected using a standardized form, to record at minimum: date, location, species, number of individuals, and state of decomposition. Five Decomposition Condition Categories (DCC) were used: 1. extremely fresh caracass (just dead); 2: fresh carcass; 3 moderate decomposition, 4: advanced decomposition; 5: mummified or skeletal remains (IJsseldijk et al., 2019). Whenever possible, the sex, dental formula (i.e. number of teeth on the lower and upper jaw) and different morphometrics measurements and tissue samples were collected and a necropsy was carried out by a veterinary from the stranding network (VE or others). *Kogia* species were distinguished based on physical characteristics and in particular on the number of teeth (*K. sima*:14–24 mandibular teeth, rarely 26 and 0–6 maxillary teeth; *K. breviceps*: 24–32 mandibular teeth, rarely 20–22 and no maxillary teeth) and the height of the dorsal fin (D) relative to body length (L) (D/L ratio > 5% in *K. sima*; <5% in *K. breviceps*; Ross, 1979). A confidence rating of "low", "medium" or "high" was assigned to the species identification (Plön et al., 2023b). When possible, identification was confirmed genetically. DNA was extracted from skin samples using DNeasy™ Tissue Kits (Qiagen) and the control region (~800 base pairs) of the mitochondrial DNA (mtDNA) was amplified by Polymerase Chain Reaction (PCR) using existing primers (Dlp-1.5 f from Baker et al., 1998 and MtCRr from Hoelzel and Green, 1998). PCR were carried out in a 25 µL reaction volume consisting of 5–50 nano grams (ng) of extracted DNA, 0.2 µM of each primer, 0.15 mM of mixed dNTPs, 1.5 mM MgCL$_2$, 2 units of SilverStar *Taq* DNA polymerase and 1x reaction buffer (Eurogentec® Inc). PCR amplifications comprised 6 min (min) at 94 °C followed by 39 cycles of 45 s (s) at 94 °C, 30 s at 50–56 °C and 50 s at 72° followed by 9 min at 72 °C. PCR amplification products were sequenced using BigDye (v. 3.1, ThermoFisher Inc.) and by electrophoresis using a 3730XL DNA Analyzer™ (ThermoFisher Inc.). Species identification was confirmed by matching the mtDNA sequences against the NCBI depository, based on 100% similarity with known *K. sima* and *K. breviceps* sequences. Strandings for which confidence in species identification was low and no genetic identification was available were referred as *Kogia sp.* For strandings reported in 2006 and before, species identification was confirmed based on photographs and by inspection of the specimens kept or reproduced to scale by the Natural History Museum of Reunion.

3. Results

3.1 Distribution surveys

A total of 2598 surveys were conducted around Reunion island over the 13 year study period representing a total distance of 101,203 km from boat-based surveys and 87,588 km from aerial surveys (Figs. 1 and 2). Surveys were generally conducted in good to excellent visibility conditions, with visibility index of 4 and 5 representing between 75% and 80% of the survey effort, respectively (Table 1, SM1). A total of 84 sightings of *Kogia* were made, which were all reported in good to excellent sea-state conditions (27% in visibility index 4 and 73% in visibility index 5). Of the 84 sightings of *Kogia*, one was a group of two pygmy sperm whales and 38 were confirmed dwarf sperm whales (*K. sima*), of which 24 were recorded during boat-based surveys (Fig. 1) and 14 during aerial surveys (Fig. 2), representing a cumulative number of 78 individuals. Forty-five sightings of *Kogia* could not be identified to species level (44 from aerial surveys and one from boat-based surveys).

Most of the sightings of *K. sima* were detected along the northern coasts of the island (50% kernel contour, Fig. 3), where systematic boat-based and aerial survey effort was completed year-round (Table 1). Three *K. sima* sightings occurred on the west coast of the island, and one in the south. No sightings

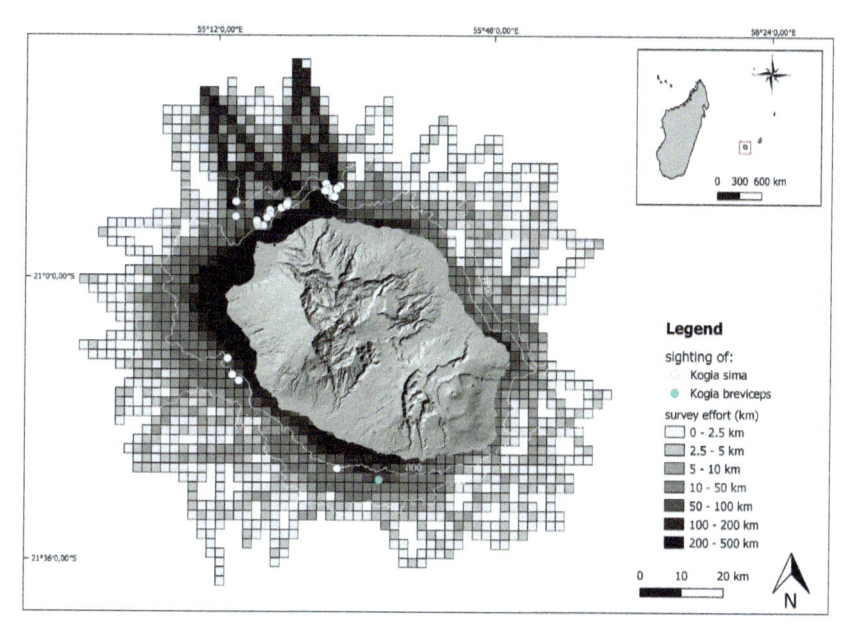

Fig. 1 Map showing boat-based survey effort conducted around Reunion between 2010 and 2023 and the location of the sightings of dwarf (*Kogia sima*) and pygmy (*Kogia breviceps*) sperm whales.

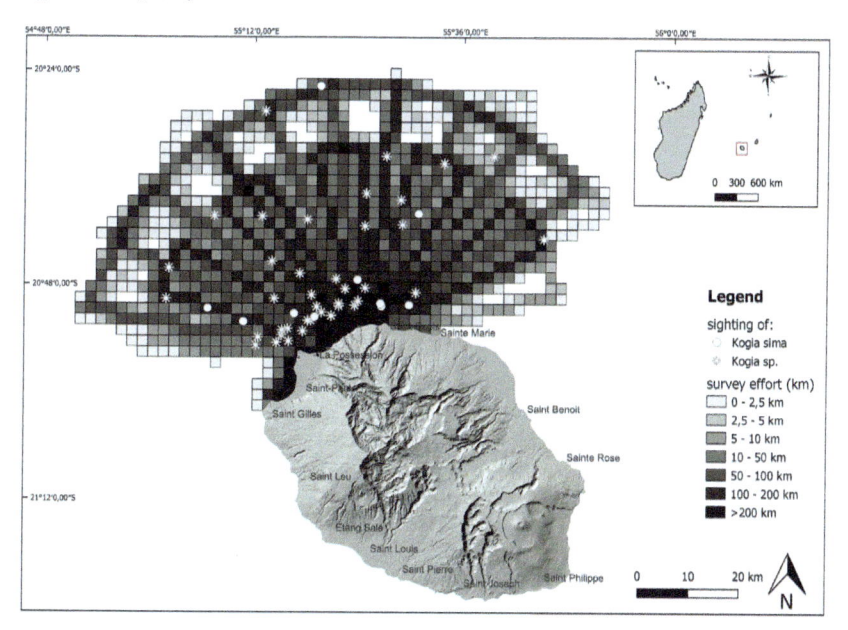

Fig. 2 Map showing aerial survey effort conducted in the northern part of Reunion between January 2015 and June 2022 and location of the sightings of dwarf sperm whales (*Kogia sima*) and *Kogia* sp.

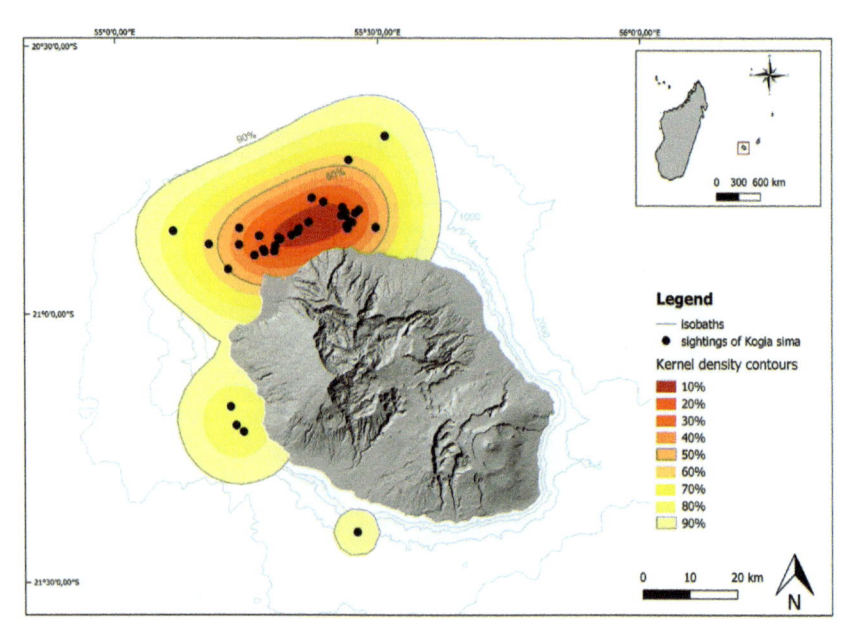

Fig. 3 Map showing the distribution of the sightings of dwarf sperm whales (*K. sima*) from boat-based and aerial surveys, and kernel density contours.

were reported from the east coast of the island, where survey effort was lower. The single sighting of *K. breviceps* occurred in the south, during boat-based survey, and kernel analysis could not be performed for this species.

Sightings of confirmed *K. sima* were mostly of single individuals (45%, n = 17), but groups of 2 to 5 individuals were also reported (Table 2). Out of the 38 sightings of *K. sima*, 7 included a mother with a calf or a juvenile. Mean group size for *K. sima* was 2.0 (SD = 1.2, min = 1, max = 5). Other sightings of *Kogia sp.*, that could not be confirmed to species, had similar group sizes (1.8, SD = 1.4, min = 1, max = 7, Kruskall-Wallis Chi-squared H = 1.4826, p = 0.223).

The majority of *K. sima* sightings occurred in waters between 800 m and 1500 m of depth (n = 28, 74%) and at a distance of 3–10 km from the coast (n = 33, 87%). The mean depth of *K. sima* sightings was 1310 m (SD = 518, min = 653, max = 2500) and mean distance to the shore was 8.2 km (SD =7.3, min = 3.2, max = 45.5, Table 2). There was no difference in depth between sightings of *K. sima* and unidentified *Kogia sp.* (Kruskall-Wallis Chi-squared H = 3.818, p = 0.051), although the latter occurred at greater distance to shore, being mainly reported from aerial surveys (Kruskall-Wallis Chi-squared H = 9.601, p = 0.002).

Table 2 Mean group size, depth and distance to the coastline of dwarf sperm whales (*K. sima*) and *Kogia sp.* sighted in January 2010–June 2022 around Reunion (all surveys combined), and values for the single sighting of pygmy sperm whale (*K. breviceps*).

	K. sima (n = 38)			*K. breviceps* (n = 1)	*Kogia* sp. (n = 45)		
	Mean	SD	Range		Mean	SD	Range
Group size	2.0	1.2	1–5	2	1.8	1.4	1–7
Bottom depth (m)	1331	518	653–2500	1000	1786	858	408–3700
Distance to shore (km)	8.2	7.3	3.2–45.5	5	15.9	11.9	2.9–48,5

Sightings of *Kogia* occurred throughout the year, with slightly higher number of *K. sima* sightings reported in October-November (n = 20) (Fig. 4). Calves were also sighted throughout the year (Fig. 4). Depth and distance to the nearest coast of *K. sima* sightings were not significantly different between months (Kruskall-Wallis Chi-squared H = 1.568, p = 0.979 and H = 6.368, p = 0.497 respectively), which was also true when combining all sightings of *Kogia* (H = 11.895, p = 0.72 and H = 13.086, p = 0.289 respectively), and no seasonal pattern was observed (Fig. 5).

3.2 Environmental DNA

The filters yielded an average of 8.15 ng/µL of DNA and no contamination was detected in the negative controls. Among the high-quality sequences produced with the MiMammal primers, a total of 459,813 reads were assigned to cetacean taxa. Detection of cetaceans occurred at 29 of the 33 sampling sites and resulted in the identification of a total of 8 species, based on a 100% similarity with referenced sequences from NCBI along the full length of the 12S rRNA amplified sequence (171 bp). Among the 8 cetacean species detected were the two species of *Kogia*. The dwarf sperm whale (*K. sima*) was identified at two sampling locations, with 8188 and 25,598 reads, respectively, matching at 100% similarity to the referenced *K. sima* sequence from the NCBI database (Accession number: NC_041303.1). The pygmy sperm whale (*K. breviceps*) was detected at one sampling location with 11,200 reads matching at 100% similarity the referenced *K. breviceps* sequence from the NCBI database (Accession number: AJ554055.1). Reads that matched at 100% similarity with *K. sima* matched at only 94.1% similarity with *K. breviceps* (and vice-versa). *K. sima* and *K. breviceps* had 10 nucleotide differences along the 171 bp sequence (94.1% similarity), supporting the discrimination at the species level. *Kogia* were not reported in the BLAST hit results from other locations.

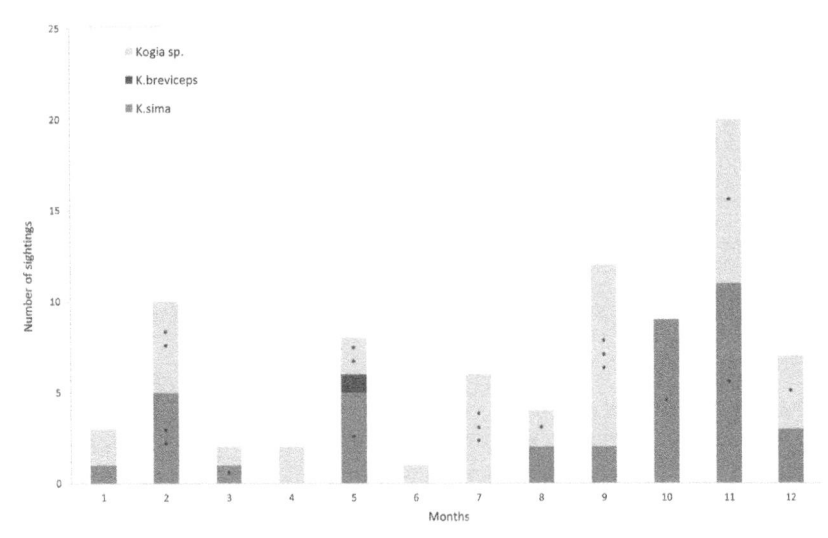

Fig. 4 Seasonal distribution of dwarf sperm whales (*K. sima*), pygmy sperm whales (*K. breviceps*) and Kogia sp. (unidentified Kogia), recorded around Reunion in 2010–2023 (all surveys combined), with *representing the number of calves.

Detections of *Kogia* from eDNA were located in the northern part of the island (Fig. 5), in waters of 800–1800 m of depth for *K. sima* and 1600 m for *K. breviceps*. The detection of *K. breviceps* occurred in February 2019, while *K. sima* was detected in samples collected in February and March 2022 (Table 3).

The other cetacean species identified with 100% similarity were the humpback whale (*Megaptera novaeangliae*), the bottlenose dolphin (*Turisops truncatus*), the Indo-Pacific bottlenose dolphin (*T. aduncus*), the pantropical spotted dolphin (*Stenella attenuata*), the melon-headed whale (*Peponocephala electra*) and Fraser's dolphin (*Lagenodelphis hosei*), with the number of reads per sample ranging between 124 and 57,914 (Table SM2.1 and Figure SM2.1 in Supp. Mat 2).

3.3 Strandings

A total of 10 Kogia were reported dead in Reunion between 1993 and 2023 (Table 4). Among them, nine were found stranded on the coast, and one was reported drifting at sea and brought back to shore in February 2022. For convenience, all events are referred to as "strandings" in the manuscript. Photographs and measurements were not available from the specimen reported in 1993 in the west (Saint Paul) so the identification

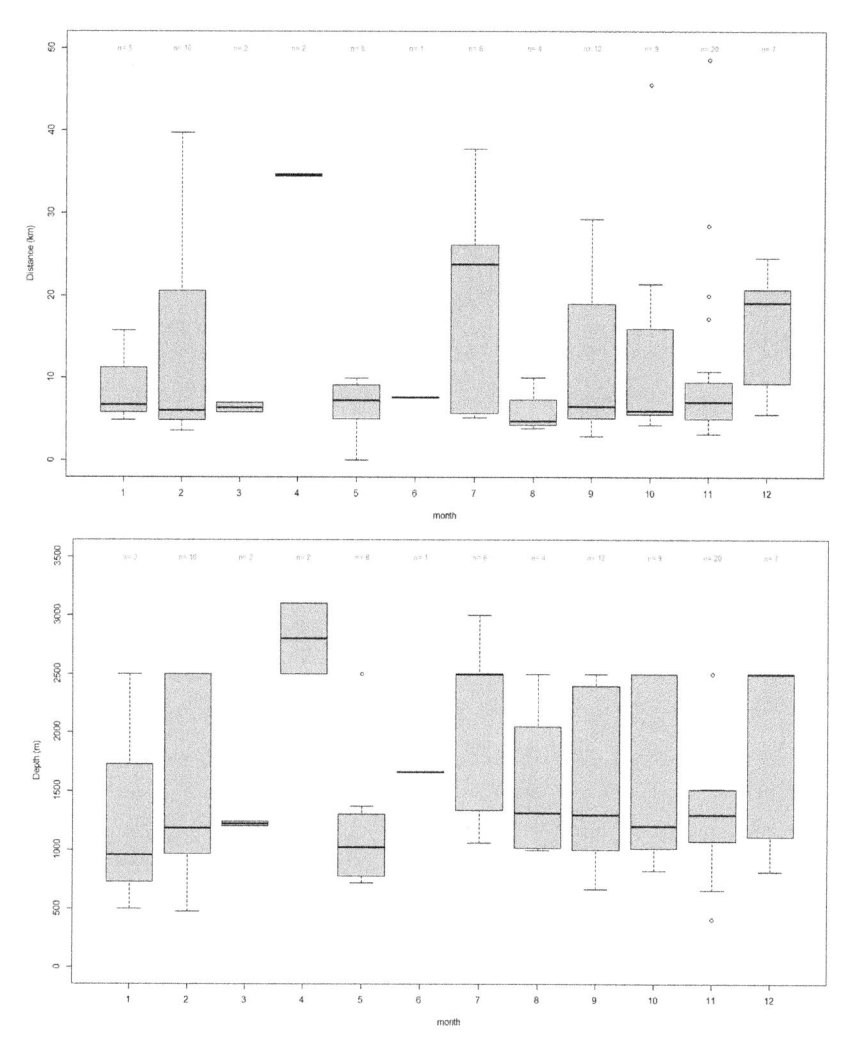

Fig. 5 Box plot of the distance to the shore and depth of *Kogia* sightings (including *K. sima*, *K. breviceps* and *Kogia sp.*) across months. Median values are represented by the middle horizontal line, with upper and lower box lines representing the 75th and 25th quartile respectively (50% of sightings are within the box), vertical lines represent 1.5 times the interquartile range, circles represent outliers, and sample size is indicated at the top.

could not be confirmed to species level. An animal that stranded in 1998 in the South of the island (Saint-Pierre) was kept at the National History Museum of Reunion and was identified *a posteriori* as a *K. breviceps*, although the specimen was in an advanced state of decomposition when

Table 3 Summary of the eDNA results presenting sampling location where *Kogia* were detected total number of reads (Nr) attributed to cetacean taxa in the sample and number of reads attributed to dwarf (*K. sima*) and pygmy (*K. breviceps*) sperm whales (and % of the number of cetacean reads) and bottom depth (in m) at sampling location.

Sampling site	Date	Latitude	Longitude	N reads cetaceans	N reads *K. sima* (%)	N reads *K. breviceps* (%)	Depth (m)
14	06/02/2019	−20.78868	55.44835	15,315	0	11,200 (73.13%)	1600
22	27/02/2022	−20.77166	55.36057	25,598	25,598 (100%)	0	1800
32	15/03/2022	−20.85961	55.65447	8188	8188 (100%)	0	800

examined (most teeth missing and decomposed dorsal fin). Individuals stranded between 2003 and 2022 (n = 7) were identified as *K. sima* with a high level of confidence after examination of the specimens on site. For three of these animals, identification was confirmed genetically (Table 4). One juvenile animal (155 cm) that stranded in 2023 was identified as a pygmy sperm whale (*K. breviceps*) with a medium level of confidence (26 teeth on the lower mandibular, dorsal fin to total length ratio of 5% (Ross, 1979), while body proportions might be biased in young individuals). Skin samples were taken for further genetic identification.

Strandings of *Kogia* were distributed all around the island (Fig. 6). One stranding of *K. sima* occurred in the south (Saint-Pierre), one in the east (Sainte-Anne), two in the west (Saint-Paul), including one collected at sea, and three in the north (Sainte-Marie). The latter strandings all occurred at the same location behind the port, within a few meters of one another, with one each in February 2006 (adult), in May 2020 (calf) and in November 2020 (pregnant female).

Necropsies were conducted on four fresh *K. sima* individuals. The individual that stranded in December 2003 in Saint Paul was a young, emaciated male, with no teeth, indicating that the animal was not yet weaned. A gross examination was carried out on the animal. About 600 mL of sero-hemorrhagic effusion was present in the abdominal cavity. Unfortunately, this fluid was not analyzed, and the heart was not dissected to look for a heart disease. The tail fluke was missing with a clear cut of the caudal peduncle. The necropsy revealed that the fluke was cut after the death of the animal.

The necropsy of the individual stranded in March 2017 in the south (Saint-Pierre) revealed that this mature male was in a good body condition with a stomach full of otoliths, fish bones and cephalopods beaks (analysis yet to be undertaken). A high level of parasitism by *Anisakis simplex* was found in the stomach and the small intestine of the animal. An abundant tracheo-bronchial foam was also present. The tympanic bullas were analyzed and did not reveal any lesions associated with exposure to underwater noise.

The individual that stranded in November 2020 in the north (Saint Marie) was a pregnant female, with a 58 cm long foetus. This animal was in a good body condition and the specimen was fresh (dermal abrasions revealed that the animal had stranded alive). A diffuse hematoma was present along the left side. The forestomach was opened and showed the presence of otoliths, providing evidence that the animal had fed shortly before its death. The thoracic tissues examination showed a pulmonary

Table 4 Details of *Kogia* strandings recorded around Reunion from 1993 to 2023.

Date	Location	Latitude	Longitude	Species	Confidence	Photo/DDC	Genetic confirmation	Sex	Age class	L	D	D/L ratio (%)	S	S/L ratio (%)	Nb of teeth	Necropsy/ Histopathologica analysis	Possible cause of death
1993	St Paul	−21.01849	55.23850	*Kogia sp.*				U	Adult							None	—
1998	St Pierre	−21.33495	55.45412	*K. breviceps*	Medium	Yes	Sample to be analysed	2[a]	Adult	203						None	—
30/12/2003	St Paul	−21.00309	55.27217	*K. sima*	High	Yes	Yes	M	Calf	118[b]	5.6	4.75	6	5.08	0/0	-Emaciated -Fluke missing (post mortem cut) −600 mL of sero-hemorrhagic effusion in the abdominal cavity	Bycatch Pathology -probable heart failure
04/02/2006	Ste Marie	−20.89393	55.54003	*K. sima*	High	Yes	No sample	2	U	Adult	230	14.4	6.26	14.5	6.30		None	—
08/05/2009	Ste Anne	−21.06894	55.73655	*K. sima*	High	Yes	Yes	4	U	Adult	231	—	—	—	—	None	—
23/03/2017	St Pierre	−21.35111	55.48559	*K. sima*	High	Yes	Yes	1	M	Adult	225	18	8.00	15	6.67	-High level of parasitism (*Anisakis simplex*) in the stomach and small intestine - Good nutritional state - Full stomach -Abundant tracheobronchial foam	Parasitic infestation

29/05/2020	Ste Marie	−20.89352	55.53850	K. sima	High	Yes	Sample to be analysed 3	U	Calf	111	5.9	5.32	8	7.21	0/10	None	—	Ship-strike/ Bycatch
21/11/2020	Ste Marie	−20.89349	55.53850	K. sima	High	Yes	Sample to be analysed 2	F	Adult	222	14	6.31	—	—	—	- Good nutritional state - Pregnant, foetus = 58 cm -diffuse hematoma on the left side -presence of otoliths in the forestomach -Pulmonary edema -abundant serohemorrhagic foam on the section of lungs and trachea -Disseminated petechiae into the melon	—	
16/02/2022	St Paul	−20.97223	55.2500	K. sima	High	Yes	Sample to be analysed 2	M	Calf	98	6	6.12	5	5.10	0/14	-Neonate -Good nutritional state -digested milk in the stomach and intestines -Numerous subcutaneous diffuse hematomas -Pulmonary	Chronic pathology Possible peri-mortem collision	

(continued)

Table 4 Details of *Kogia* strandings recorded around Reunion from 1993 to 2023. *(cont'd)*

Date	Location	Latitude	Longitude	Species	Confide-nce	Photo	Genetic confirmation	DDC	Sex	Age class	L	D	D/L ratio (%)	S	S/L ratio (%)	Nb of teeth	Necropsy/ Histopathological analysis	Possible cause of death
0/09/2023	Trou d'eau	−21.10168	55.24499	*K. breviceps*	Medium	Yes	Sample to be analysed	4	F	Juvenile	155	7	4.52	16.5	10.65	0/26	None	—

Necropsy/Histopathological analysis: edema -Abundant sero-hemorrhagic tracheobronchial foam -Histopathologica analysis: prominent alveolar histiocytosis suggesting a chronical process consistent with heart or respiratory failure

DCC: Decomposition condition Category (1: very fresh; 2: fresh carcass; 3: moderate decomposition, 4: advanced decomposition, U: unknown), sex (Male/Female/Unknown), Age class, total length (L, in cm), dorsal fin height (D, in cm), Snout-to-blow hole distance (S, in cm), S/L ratio, necropsy comments, and possible cause of death.

[a]DCC: 2 when stranded, DCC:4 when examined;
[b]Based on the reconstruction of the individual by the Natural History Museum.

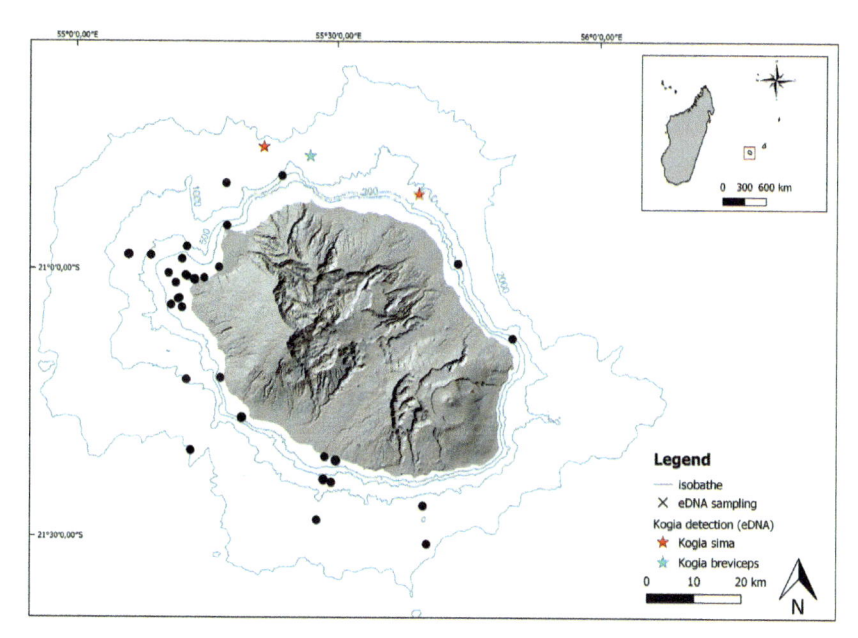

Fig. 6 Map showing the distribution of the eDNA sampling locations and the detection of K. sima and K.breviceps around Reunion Island.

edema, with an abundant sero-haemorrhagic foam. The sagittal section of the melon revealed disseminated petechiae.

The individual examined in 2022 was collected at sea off Saint-Paul and frozen until the necropsy could be carried out. The necropsy revealed that it was a neonate, in good body condition. Numerous subcutaneous diffuse hematomas were present (in the ventral part and around the head on the right and left side of the body). Sections of these hematomas showed that they were made before or at the time of the death. Inspection of the digestive tract showed the presence of digested milk in the stomach and the intestines. The lungs were enlarged, with hemorrhagic foci disseminated along the tissues and pulmonary edema. An abundant sero-haemorrhagic tracheobronchial foam was present when trachea and lungs were cross-sectioned.

4. Discussion

The combination of sighting data, eDNA detections and stranding events confirmed that both species of *Kogia* are present within the territorial waters (<22 km from shore) of Reunion and might possibly be sympatric. The dwarf sperm whale, *K. sima,* appeared to be more common than the pygmy sperm whale, *K. breviceps.* The northern part of the island might provide suitable habitats for the dwarf sperm whale relatively close to shore while the examination of stranded individuals highlighted potential threats to the species arising from human activities.

4.1 Species occurrence

Most *Kogia* detected during boat-based surveys were identified as dwarf sperm whales (24 sightings of *K. sima,* and one sighting of undetermined *Kogia* sp.). Only one confirmed sighting of pygmy sperm whales was reported during the boat-based surveys and none from the aerial surveys. Strandings data also mirror this trend, with seven strandings of dwarf sperm whales recorded over the last decade, compared to two strandings of pygmy sperm whales in 1998 and 2023.

Both dwarf and pygmy sperm whales were detected in the eDNA sequences, confirming the use of Reunion waters by both species. Detecting the two species in the relatively low number of sampling stations (n = 33) at the surface was particularly remarkable, and further emphasizes the efficiency of this emergent technique to detect cryptic, deep-diving species like the dwarf and pygmy sperm whales, which have a low probability of detection in traditional visual surveys (Juhel et al., 2021).

The combination of records from different sources indicates that the dwarf sperm whale is more common in the territorial waters of Reunion than initially reported, while also confirming the presence of the pygmy sperm whale (Dulau-Drouot et al., 2008). These results are consistent with dwarf sperm whales being more common in tropical habitats than pygmy sperm whales, which prefer more temperate waters (Bloodworth and Odell, 2008; Caldwell and Caldwell, 1989; Moura et al., 2016). In the western tropical Indian Ocean, the dwarf sperm whales appears to be relatively common in the Maldives, a chain of island located close to the equator (0–7°N of latitude), with 74 sightings reported (representing 4.2% of all on-effort sightings), while the pygmy sperm whale has not been recorded (Anderson, 2005). This is consistent with the results of a wide-ranging vessel-based survey conducted in the western Indian Ocean, which indicated that dwarf sperm whale was relatively frequent north of the Seychelles, in waters between 5°S and 7°N of latitude (Ballance and Pitman, 1998). Large-scale line-transect aerial surveys also confirmed a high density of *Kogia* around the Seychelles (Laran et al., 2017), but identification to species level could not be undertaken. In more temperate waters, such as in South Africa, where the highest numbers of *Kogia* strandings are reported from the region, both species are recorded evenly, with 106 strandings of *K. breviceps* and 85 of *K. sima* reported over a 15-year period (1880–1995) along an approximate 3000 km of coastline (Plön, 2004; Ross, 1979).

Elsewhere in the western Indian Ocean, confirmed records of both species exist, but are too sporadic to further assess any latitudinal gradient in species distribution. A recent compilation of strandings data collected over the past 20 years (2000–2020) in the south-western Indian Ocean reported 24 *Kogia* stranding events, including 11 dwarf and 9 pygmy sperm whales (Plön et al., 2023b). From these data, the highest occurrence of dwarf sperm whale strandings were reported for Reunion (n = 6), with stranding events also reported for South Africa (n = 3), Mozambique (n = 1) and Seychelles (n = 1) during that period. Pygmy sperm whale strandings were recorded along the eastern coast of mainland Africa, in South Africa (n = 4), Mozambique (n = 1) and Kenya (n = 3) and in the continental island of Madagascar (n = 1). One at-sea sighting of *K. sima* was recorded off the south-west coast of Madagascar (Cerchio et al., 2022) and sightings of both species of *Kogia* (n = 2 *K. sima* and n = 1 *K. breviceps*) have also been reported around Mayotte, in the northern Mozambique channel (Kiszka et al., 2010).

In the northern part of the western Indian Ocean, sightings, strandings and osteological records have confirmed the presence of dwarf sperm whales off Eritrea, Oman, Pakistan and India (Baldwin et al., 1999; Collins et al., 2002; Gore et al., 2012; Kumaran, 2002; Notarbartolo di Sciara et al., 2017), while the pygmy sperm whale has only been reported off Pakistan (Gore et al., 2012).

4.2 Spatial distribution around Reunion

Several datasets provide evidence of the use of Reunion territorial waters (<22 km) by the dwarf sperm whale, especially off the northern coast, where systematic line-transect surveys (boat-based and aerial) were conducted over an 8-year period. Because effort was not homogeneously distributed around the island, it is still unclear if this area represents a preferred habitat for the species, compared to other areas around the island. The western side of the island was covered relatively intensively, but effort was lower in the species preferred depth range (800–1500 m) as deeper waters occur further off-shore in this area. The few *Kogia* sightings recorded around the island were located in areas where the 800 m depth contour runs closer to the coast. Increasing effort in the 800–1500 m depth waters (>10 km off-shore) might result in more detections around the island. Furthermore, given the elusive behavior of the species at the surface, the number of sightings is likely influenced by sea state and the type of survey vessel, with observers being more likely to detect *Kogia* from a higher survey platform (as used during the systematic line-transect surveys), than from smaller boats. Line-transect surveys on a larger boat were also conducted around the island, but at a lower frequency (one four-day survey per year on average) than in the north (three one-day surveys a month) and resulted in only one sighting of dwarf sperm whale in the north. Barlow (2015) reported that in Beaufort 2 sea conditions the probability of sighting *Kogia* during large vessel surveys drops to less than 10% of their sighting probability in Beaufort 0 conditions. In this study, visibility was generally good to excellent during surveys (representing from 75% to 90% of the survey effort) and sightings of *Kogia* were only made in visibility index of 4 and 5. Future studies for the species should consider the effect of survey effort (including sighting conditions and platform type) and environmental variables, such as depth, to further assess spatial distribution around the island and identify preferred habitats.

Results from eDNA sampling and stranding events also indicate more detections in the north, although the numbers were too low to be conclusive. Out of the 33 sampling sites and eight cetacean species detected around the island, *Kogia* were only detected in the north, further supporting the use of this area by both species. *Kogia* strandings occurred all around the island, but three of them (i.e., 33%) were found in the same location in the north, behind the port of Sainte-Marie, within a few meters of one another. The carcasses of two adults were fresh (in February 2006 and in November 2020), while the calf showed advanced decomposition (May 2020). Given that the access to the beach was closed due to the renovation of the port in early 2020, the discovery of the latter animal might have been delayed. The higher occurrence of strandings in the north, with most individuals discovered in a fresh decomposition state, is suggestive of individuals that were inhabiting adjacent waters.

The high number of sightings, stranding events and eDNA detections indicates that at least the northern sector of Reunion Island provides a suitable habitat for *Kogia*, and the dwarf sperm whale in particular. Evidence of site fidelity has been reported off Hawai'i, an island of similar size based on multi-year resightings and individual spatial distribution

(Baird et al., 2021). Further photo-identification work should be undertaken to assess the level of residency and the spatial ranges of individuals sighted off the northern coast of Reunion.

4.3 Habitat use

Although *Kogia* were sighted up to 48 km offshore, most sightings were distributed along the insular slope, in water depths from 800 m to 1500 m. Dwarf sperm whales occurred on average in 1310 m deep waters (SD = 518, min = 653, max = 2500) and unidentified *Kogia sp.* were reported in similar water depths (1786 m on average, SD = 858, min = 408, max = 3700). Most sightings of dwarf sperm whales off Reunion were located relatively close to the coast, at a mean distance of 8.2 km (SD = 7.3, min = 3.2, max = 45.5), and this is likely linked to the narrow shelf and steep relief around the island. It is however likely that they also use waters further off-shore as indicated by the occurrence of unidentified *Kogia sp.* from aerial surveys (mean distance from shore = 15.9 km, SD = 11.9, min = 2.9, max = 48.5). This spatial distribution is consistent with described habitat preferences for this species in other areas, although dwarf sperm whales were also recorded from oceanic waters elsewhere in the western Indian Ocean (Ballance and Pitman, 1998). Within the archipelago of Hawai'i, the mean depth of dwarf sperm whale sightings was 1425 m (SD = 954 m), with a range from 450 to 3200 m (Baird, 2005), and off the island of Hawai'i, sighting rates were highest in depths between 500 and 1000 m, and were lower than expected in shallower and deeper waters outside of this range (Baird et al., 2021). Around the Maldives, the species observed just outside the reef (Anderson, 2005). Off Great Abaco Island, in the Bahamas, dwarf sperm whales were always found in waters deeper than 300 m and were primarily distributed along the upper slope during winter, suggesting seasonal variability in their habitat preferences (Dunphy-Daly et al., 2008).

In this study, sightings of *Kogia* seemed to increase from September to November, during the austral summer. Although distribution surveys (boat-based and aerial) off the northern part of the island were conducted year-round in a systematic manner, further analyses accounting for effort and sea state would be needed to confirm seasonal trends. It would be reasonable to assume that for such a cryptic species the increased number of sightings in this period could reflect seasonal changes in weather and sea-state conditions. From eDNA sampling, *Kogia* were also detected during the austral summer, in the months of February and March. However, the sampling design did not allow for spatio-temporal comparison, which would require re-sampling each sampling site on a seasonal basis. No seasonal trends could be observed from the stranding events, which occurred throughout the year. Therefore, although the distribution data would tend to suggest an increased occurrence of *Kogia* during the austral summer, eDNA and stranding data were insufficient to confirm patterns of seasonality. Some evidence of seasonal movement has been documented off the Bahamas, with dwarf sperm whales mainly sighted during winter over slope habitats, while in summer, sighting rates decrease significantly in this stratum, together with group size (Dunphy-Daly et al., 2008). The authors suggest that this might reflect inshore-offshore movement, although it could also indicate a seasonal shift in distribution. Elsewhere, there is little evidence of seasonality, although seasonal differences in dwarf sperm whale stranding records have been observed in some areas that could reflect seasonal movements (McAlpine, 2009; Moura et al., 2016).

4.4 Group size and biology

Group sizes of dwarf sperm whales observed around Reunion ranged from 1 to 5 (mean = 2.0, SD = 1.2), with 46% of the sightings being of single individuals and the only sighting of pygmy sperm whales included 2 individuals. These results are consistent with the relatively small group sizes reported from other areas, with solitary animals or cow/calf pairs most commonly reported (Anderson, 2005; Baird et al., 2021; Ballance and Pitman, 1998; Dunphy-Daly et al., 2008; Ross, 1984). Around Reunion, sightings of calves were relatively common (n = 17) and occurred in groups of 3–5 individuals (except for a single mother-calf pair), suggesting some level of social grouping. However it remains to be investigated whether their social organization mirrors that of the common sperm whales, with females with their young forming social units (Best, 1979; Whitehead, 2003). Strandings data from South Africa tend to indicate that some age/sex segregation may occur, with groups consisting of solitary adult animals of both sexes, cow/calf pairs and small groups of immature animals (Plön and Baird, 2022; Ross, 1984).

Calves were observed in different months but the number of sightings were too few to provide any insights on breeding seasonality. However, the size of the foetus (58 cm) of the lone pregnant female examined during this study was consistent with the growth curve (fetal and calf body lengths plotted against date) established from stranding records of the dwarf sperm whale from the southern hemisphere (Pinedo, 1987) and indicated that birth would have occurred during the austral summer (Fig. 6). This estimate was also consistent with stranding data from South Africa, that indicates that births (and conception) occur between December and March (Plön, 2004). The month of stranding versus body size of calves reported in Reunion were also consistent with the southern hemisphere growth curve of *K. sima* from Pinedo (1987) and support a birthing period during the austral summer (Fig. 7). That this is not reflected in the sightings data might be linked to the difficulty of discriminating calves from juveniles at sea. The three *K. sima* calves that stranded in Reunion ranged from 98 to 118 cm in length which is consistent with the estimated body length at birth of approximately 1 m (Ross, 1979; Pinedo, 1987), although the 98 cm calf stranded in Reunion seem to be the smallest newborn individual recorded from the region (Plön, 2004; Ross, 1979). The necropsy indicated the presence of milk in the stomach, indicating that it was a neonate and not stillborn. The young *K. breviceps* stranded in September 2023 was 115 cm long, while length at birth has been estimated to be approximately 120 cm for this species (Ross, 1979). The species identification was based on the D/L ratio (4.5) and the number of teeth (26), but should be further confirmed genetically Fig. 8.

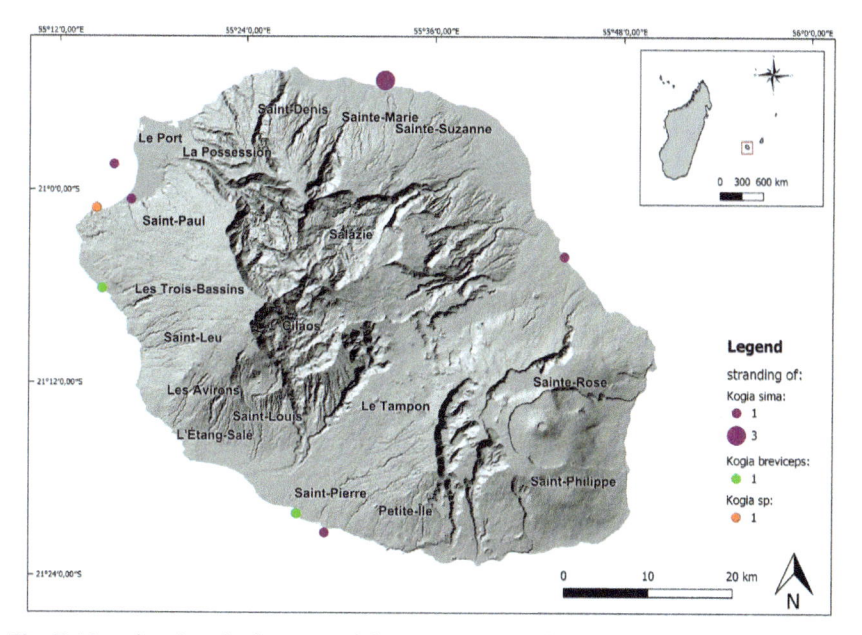

Fig. 7 Map showing the location of the 9 strandings of *K. sima*, *K. breviceps* and *Kogia sp.* around Reunion in 1993-2022.

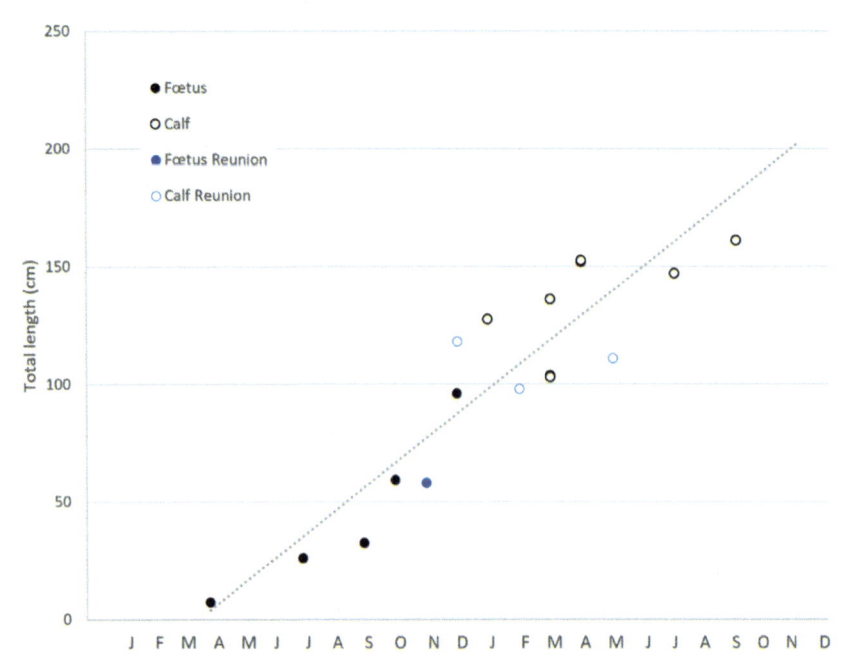

Fig. 8 Growth curve previously established by Pindeo (1987) for southern hemisphere *K. sima*, reporting the total length (in cm) of foetus versus month of foetus and calves, with added values for Reunion strandings.

4.5 Threats

Of the four fresh animals that were necropsied, three were in good body condition/nutritional state, with evidence of recent feeding, indicating that these animals were healthy enough to catch prey or to suckle milk shortly before they died. These findings, together with the results of the necropsies, ruled out any severe chronic causes of death, and are rather in favor of an acute event. The presence of an abundant sero-hemorrhagic tracheo-bronchial foam demonstrated that three animals died from hypoxia, inducing death. Indeed, hypoxia induces a loss of membrane integrity in the lung tissue, allowing fluid to leak into the airways. The fluid is then combined to residual air and forms the foam (Davis and Bowerman, 1990). For two animals, the thoracic tissues examination showed a pulmonary edema and foam was present in the lungs, indicating a possible drowning. Moreover, the presence of diffuse subcutaneous hematomas on the side or the ventral part of the body of these two animals is consistent with a mechanical trauma. For the female stranded in November 2020, disseminated petechiae were present in the melon, that would corroborate the trauma. All these findings (i.e., exclusion of other causes of death, persistent froth, oedematous lungs and bruises) have been described in cases of bycatch or ship strike (Kuiken, 1994). Hence, even if the exact cause of death for these two animals could not be ascertained, they are likely linked to anthropogenic activities. Reunion is located on one of the major shipping lanes of the south-western Indian Ocean, within which most of the marine traffic transiting between South Africa and Asia is concentrated (Tournadre, 2014) and the main commercial port is located in the north of the island, where relatively high concentrations of *Kogia* were reported.

5. Conclusions

In conclusion, this study utilized data from a variety of research approaches, including stranding reports, visual aerial and boat-based surveys, and environmental DNA techniques in order to provide new insights on the occurrence and distribution of *Kogia* in the waters of Reunion. The dwarf sperm whale, *K. sima*, was the most frequent species identified at sea and reported in stranding data. Comparatively, the pygmy sperm whale, *K. breviceps*, seemed to be less common, although uncertainty remains for some sightings which could not be identified to species level. Both species were detected in the eDNA water samples collected in the north of the island, confirming that dwarf and pygmy sperm whales are likely sympatric in Reunion waters, although habitat preferences could not be further clarified. Moreover, the reported strandings highlighted the importance of considering the vulnerability of these species to human activities. The confirmed high use of at least the northern part of the island serves as a scientific baseline for future research aimed at further investigating habitat use and suggests the need for further investigations into potential threats and conservation measures.

Furthermore, this multi-disciplinary study on *Kogia* highlighted the complementarity and limits of the different research approaches. The study confirmed the efficiency of environmental DNA in detecting and discriminating between the two *Kogia* species. This technique could be used in future surveys to further investigate seasonality. Visual surveys, while requiring substantial effort, were deemed effective under optimal sighting conditions, which should be considered when comparing levels of occurrence with other parts of the world where *Kogia* are reported to be relatively common. Aerial surveys were efficient for covering large areas but posed challenges in species identification. While strandings provided the first records of Kogia in Reunion, and are the only source of available information in some areas, the necropsies of stranded animals represented highly valuable resources for inferring threats to these cryptic species. Acoustic data were not investigated, but hydrophones were deployed around the island as a continuation of this study to enhance knowledge of *Kogia* occurrence and seasonality patterns.

Acknowledgments

Vessel-based and aerial surveys conducted by GLOBICE and BIOTOPE in the northern part of Reunion were funded by Direction d'Opération de la nouvelle Route du Littoral (DORL), Region Reunion as part of the construction of the NRL highway. Distribution surveys around Reunion were funded by the European Union (FEDER/INTEREG), DEAL-Reunion and Region Reunion, as part of different research programmes conducted by Globice (CeTO, EtCETRA, COMBAVA, DECLIC). Environmental DNA study was funded by DEAL-Reunion and the European Union (FEDER, INTERREG, program DECLIC, and authorized under the Access to Genetic Resources and Benefit Sharing permit: TREL2118715S/542). Stranding data were collected by the local stranding network, coordinated by Globice, as part of the National Stranding Network coordinated by Observatoire Pelagis, France, with a funding from DEAL-Reunion. We are grateful to the Museum of Natural History of Reunion for providing access to the specimens reproduced by Salim Isaach (from animals stranded in 1998 and 2003), to Robin Baird, from Cascadia Research Collective, for his help in species identification and useful comments on the manuscript, to Tim Collins for editing the English, to Globice volunteers for their participation in boat-based surveys around the island and to the anonymous reviewers who helped improve the manuscript.

Appendix A. Supporting information

Supplementary data associated with this article can be found in the online version at https://doi.org/10.1016/bs.amb.2024.08.003.

References

Anderson, C., 2005. Observations of cetaceans in the Maldives, 1990–2002. J. Cetacean Res. Manage. 7 (2), 119–135.

Arnason, U., Gullberg, A., Janke, A., 2004. Mitogenomic analyses provide new insights into cetacean origin and evolution. Gene 333, 27–34. https://doi.org/10.1016/j.gene.2004.02.010.

Baird, R.W., 2005. Sightings of dwarf (*Kogia sima*) and pygmy (*Kogia breviceps*) sperm whales from the main Hawaiian Islands. Pac. Sci. 59, 461–466. https://doi.org/10.1353/psc.2005.0031.

Baird, R.W., Mahaffy, S.D., Lerma, J.K., 2021. Site fidelity, spatial use, and behavior of dwarf sperm whales in Hawaiian waters: using small-boat surveys, photoidentification, and unmanned aerial systems to study a difficult-to-study species. Mar. Mamm. Sci. 3, 326–348. https://doi.org/10.1111/mms.12861.

Baker, C.S., Florez-Gonzalez, L., Abernethy, B., Rosenbaum, H.C., Slade, R.X., Capella, J., et al., 1998. Mitochondrial DNA variation and maternal gene flow among humpback whales of the southern hemisphere. Mar. Mamm. Sci. 14, 721–737.

Baldwin, R.M., Gallagher, M., Van Waerebeek, K., 1999. A review of cetaceans from waters off the Arabian peninsula. In: Fisher, M., Ghazanfar, S.A., Spalton, J.A. (Eds.), The Natural History of Oman: A Festschrift for Michael Gallagher. Backhuys Publishers, Leiden, pp. 161–189.

Ballance, L.T., Pitman, R.L., 1998. Cetaceans of the western tropical Indian Ocean: distribution, relative abundance, and comparisons with cetacean communities of two other tropical ecosystems. Mar. Mamm. Sci. 14 (3), 429–459.

Barlow, J., 2006. Cetacean abundance in Hawaiian waters estimated from a summer/fall survey in 2002. Mar. Mamm. Sci. 22 (2), 446–464.

Barlow, J., 2015. Inferring trackline detection probabilities, g(0), for cetaceans from apparent densities in different survey conditions. Mar. Mamm. Sci. 31, 923–943. https://doi.org/10.1111/mms.12205.

Best, P.B., 1979. Social organization in sperm whales, *Physeter macrocephalus*. In: Winn, H.E., Olla, B.L. (Eds.), Behavior of Marine Animals. Plenum, New York, pp. 227–289.

Blainville, M.H., 1838. Sur les Cachalot [On the Sperm Whales]. Annales Françaises 2, 335–337.

Bloodworth, B.E., Odell, D.K., 2008. *Kogia breviceps* (Cetacea: Kogiidae). Mamm. Species 819, 1–12. https://doi.org/10.1644/819.1.

Caldwell, D.K., Caldwell, M.C., 1989. Pygmy sperm whale *Kogia breviceps* (de Blainville, 1838): dwarf sperm whale *Kogia simus* (Owen, 1866). In: Ridgway, S.H., Harrison, S.R. (Eds.), Handbook of Marine Mammals: River Dolphins and the Larger Toothed Whales. Academic Press, London, pp. 235–260.

Cerchio, S., Laran, S., Andrianarivelo, N., Saloma, A., Andrianantenaina, B., Van Canneyt, O., et al., 2022. Cetacean species diversity in Malagassy waters. In: Goodman, S.M. (Ed.), The New Natural History of Madagascar. Princeton University Press, Princeton, pp. 411–424.

Chivers, S.J., Leduc, R.G., Robertson, K.M., Barros, N.B., Dizon, A.E., 2006. Genetic variation of Kogia spp. with preliminary evidence for two species of Kogia sima. Mar. Mamm. Sci 21 (4), 619–634.

Clarke, M.R., 2003. Production and control of sound by the small sperm whales, *Kogia breviceps* and *K. sima* and their implications for other cetacea. J. Mar. Biol. Assoc. U. Kingd. 83 (2), 241–263. https://doi.org/10.1017/S0025315403007045h.

Collins, T., Minton, G., Baldwin, R.L., Waerebeek, K.V., Davies, A., Cockcroft, V.G., 2002. A preliminary assessment of the frequency, distribution and causes of mortality of beach cast cetaceans in the Sultanate of Oman, January 1999 to February 2002. Scientific Committee document SC/54/O4, International Whaling Commission, 26 April–10 May 2002, Shimonoseki, Japan.

Davis, J.H., Bowerman, D.L., 1990. Bodies found in water. Handbook of Forensic Pathology, Froede, R.C. College of American Pathologists, Northfield, Illinois, pp. 140–147.

Dulau-Drouot, V., Boucaud, V., Rota, B., 2008. Cetacean diversity off La Réunion island (France). J. Mar. Biol. Assoc. UK 88 (6), 1263–1272. https://doi.org/10.1017/S0025315408001069.

Dunphy-Daly, M.M., Heithaus, M.R., Claridge, D.E., 2008. Temporal variation in dwarf sperm whale (*Kogia* sima) habitat use and group size off Great Abaco Island, Bahamas. Mar. Mamm. Sci. 24, 171–182. https://doi.org/10.1111/j.1748-7692.2007.00183.x.

Garrison, L.P., Martinez, A., Maze-Foley, K., 2010. Habitat and abundance of cetaceans in Atlantic Ocean continental slope waters off the eastern USA. J. Cetacean Res. Manag. 11, 267–277.

Gore, M.A., Kiani, M.S., Ahmad, E., Hussain, B., Ormond, R.F., Siddiqui, J., et al., 2012. Occurrence of whales and dolphins in Pakistan with reference to fishers' knowledge and impacts. J. Cetacean Res. Manag. 12, 235–247.

Handley, C.O., 1966. A synopsis of the genus *Kogia* (pygmy sperm whales). In: Norris, K. (Ed.), Whales, Dolphins, and Porpoises. University of California Press, Berkeley, CA, pp. 63–69.

Hildebrand, J.A., Frasier, K.E., Baumann-Pickering, S., Wiggins, S.M., Merkens, K.P., Garrison, L.P., et al., 2019. Assessing seasonality and density from passive acoustic monitoring of signals presumed to be from pygmy and dwarf sperm whales in the Gulf of Mexico. Front. Mar. Sci. 6. pp. 66. https://doi.org/10.3389/fmars.2019.0006.

Hodge, L.E., Baumann-Pickering, S., Hildebrand, J.A., Bell, J.T., Cummings, E.W., Foley, H.J., et al., 2018. Heard but not seen: occurrence of *Kogia* spp. along the western North Atlantic shelf break. Mar. Mamm. Sci. 34, 1141–1153.

Hoelzel, A.R., Green, A., 1998. PCR protocols and population analysis by direct DNA sequencing and PCR-based fingerprinting. In: Hoelzel, A.R. (Ed.), Molecular Genetic Analysis of Population, A Practical Approach, second ed. Oxford University Press, Oxford, pp. 159–189.

IJsseldijk, L.L., Brownlow, A.C., Mazzariol S., 2019. Best practice on Cetacean Post Mortem investigation and tissue sampling. Information document ASCONBANS/AC25/Rev1. 25th meeting of the advisory Committee, Stralsund, Germany, 17–19 Sept. 2019, 70pp. Available at: https://www.ascobans.org/en/document/best-practice-cetacean-post-mortem-investigation-and-tissue-sampling.

Juhel, J.B., Marques, V., Polanco-Fernández, A., Borrero-Pérez, G.H., Mutis-Martinezguerra, M., Valentini, A., et al., 2021. Detection of the elusive Dwarf sperm whale (*Kogia sima*) using environmental DNA at Malpelo island (Eastern Pacific, Colombia). Ecol. Evol. 11 (7), 2956–2962. https://doi.org/10.1002/ece3.7057.

Kiszka, J., Braulik, G., 2020a. Kogia sima. The IUCN Red List of Threatened Species 2020: e.T11048A50359330. https://dx.doi.org/10.2305/IUCN.UK.2020-2.RLTS.T11048A50359330.en.

Kiszka, J., Ersts, P.J., Ridoux, V., 2010. Structure of a toothed cetacean community around a tropical island (Mayotte, Mozambique Channel). Afr. J. Mar. Sci. 32 (3), 543–551.

Kiszka, J., Braulik, G., 2020b. Kogia breviceps. The IUCN Red List of Threatened Species 2020: e.T11047A50358334. https://dx.doi.org/10.2305/IUCN.UK.2020-2.RLTS.T11047A50358334.en.

Kuiken, T., 1994. Review of the criteria for the diagnosis of by-catch in cetaceans. In: Diagnosis of By-catch in cetaceans. Proceedings of the Second European Cetacean Society Workshop on cetacean pathology. Montpellier, France, 2 March 1994, pp. 38–43.

Kumaran, P.L., 2002. Marine mammal research in India—a review and critique of the methods. Curr. Sci. 83 (10), 1210–1220.

Laran, S., Authier, M., Van Canneyt, O., Doremus, G., Watremez, P., Ridoux, V., 2017. A comprehensive survey of pelagic megafauna: their distribution, densities, and taxonomic richness in the tropical Southwest Indian Ocean. Front. Mar. Sci. 4, 139.

Madsen, P.T., Carder, D.A., Bedholm, K., Ridgway, S.H., 2005. Porpoise clicks from a sperm whale nose—convergent evolution of 130 kHz pulses in toothed whale sonars? Bioacoustics 15, 195–206. https://doi.org/10.1080/09524622.2005.9753547.

Malinka, C.E., Tonnesen, P., Dunn, C.A., Claridge, D.E., Gridley, T., Elwen, S.H., et al., 2021. Echolocation click parameters and biosonar behaviour of the dwarf sperm whale (*Kogia* sima). J. Exp. Biol. 224 (6), jeb240689. https://doi.org/10.1242/jeb.240689.

Marten, K., 2000. Ultrasonic analysis of pygmy sperm whale (*Kogia breviceps*) and Hubbs' beaked whale (Mesoplodon carlhubbsi) clicks. Aquat. Mamm. 26, 45–48.

May-Collado, L., Agnarsson, I., 2006. Cytochrome b and Bayesian inference of whale phylogeny. Mol. Phylogenet. Evol. 38 (2), 344–354.

McAlpine, D.F., 2009. Pygmy and dwarf sperm whales. In: Perrin, W.F., Würsig, B., Thewissen, J.G.M. (Eds.), Encyclopedia of Marine Mammals. Elsevier Academic Press, California, pp. 936–938.

McCullough, J.L.K., Wren, J.L.K., Oleson, E.M., Allen, A.N., Siders, Z.A., Norris, E.S., 2021. An Acoustic Survey of Beaked Whales and *Kogia* spp. in the Mariana Archipelago using drifting recorders. Front. Mar. Sci. 8, 664292. https://doi.org/10.3389/fmars.

Merkens, K., Mann, D., Janik, V.M., Claridge, D., Hill, M., 2018. Clicks of dwarf sperm whales (*Kogia* sima). Mar. Mamm. Sci. 34, 963–978.

Moura, J.F., Acevedo-Trejos, E., Tavares, D.C., Meirelles, A.C., Silva, C.P., Oliveira, L.R., et al., 2016. Stranding events of *Kogia* whales along the Brazilian coast. PLoS One 11 (1), e0146108.

Mullin, K.D., Fulling, G.L., 2004. Abundance of cetaceans in the oceanic northern Gulf of Mexico, 1996–2001. Mar. Mamm. Sci. 20 (4), 787–807.

Nikolic, N., Dulau, V., Hoarau, L., Crochelet, E., Pinfield, R., Corse, E., Tang, C.Q. submitted. Monitoring marine mammals around Reunion island (southwest Indian Ocean) using environmental DNA (eDNA) comparing to visual monitoring.

Notarbartolo di Sciara, G., Kerem, D., Smeenk, C., Rudolph, P., Cesario, A., Costa, M., et al., 2017. Cetaceans of the Red Sea. CMS Tech. Ser. 33, 86.

Owen, R., 1866. On some Indian cetacea collected by Walter Elliot, Esq. Trans. Zool. Soc. London. 6 (1), 17–47.

Pinedo, M.C., 1987. First record of a dwarf sperm whale from southwest Atlantic, with reference to osteology, food habits and reproduction. Sci. Rep. Whales Res. Inst. 38, 171–186.

Plön, S., 2004. The Status and Natural History of Pygmy (Kogia breviceps) and Dwarf (*K. sima*) Sperm Whales Off Southern Africa (PhD thesis). Department of Zoology and Entomology, Rhodes University, Grahamstown, p. 553.

Plön, S., Baird, R.W., 2022. Dwarf Sperm Whale, *Kogia* sima (Owen, 1866). In: Hackländer, K., Zachos, F.E. (Eds.), Handbook of the Mammals of Europe, Handbook of the Mammals of Europe. Springer Nature Switzerland AG 2022. https://doi.org/10.1007/978-3-319-65038-8_91-1.

Plön, S., Best, P., Duignan, P., Lavery, S.D., Bernard, R.T.F., et al., 2023a. Population structure of pygmy (Kogia breviceps) and dwarf (Kogia sima) sperm whales in the Southern Hemispher may reflect foraging ecology and dispersal patterns. Adv. Mar. Biol. 96. https://doi.org/10.1016/bs.amb.2023.09.001.

Plön, S., Norman, S., Adam, P.A., Andrianarivelo, N., Bachoo, S., Braulik, G., et al., 2023b. Spatio-temporal trends in cetacean strandings and response in the south-western Indian Ocean—2000 to 2020. J. Cetacean. Res. Manage. 24, 95–119.

Rice, D.W., 1998. Marine Mammals of the World—Systematics and Distribution. Society for Marine Mammalogy, Lawrence.

Ridgway, S.H., Carder, D.A., 2001. Assessing hearing and sound production in cetaceans not available for behavioral audiograms: experiences with sperm, pygmy sperm, and gray whales. Aquat. Mamm. 27, 267–276.

Ross, G.J.B., 1979. Records of pygmy and dwarf sperm whales. genus *Kogia*. Fromsouthern Africa. with biological notes and some comparisons. Ann. Cape Prov. Mus. Nat. Hist. 15, 259–327.

Ross, G.J.B., 1984. The smaller cetaceans of the south east coast of Southern Africa. Ann. Cape Prov. Mus. (Nat. Hist.) 15 (2), 173–410.

Shan, T., Tian, L., Liu, Y., 2019. The complete mitochondrial genome of the dwarf sperm whale Kogia sima (Cetacea: Kogiidae). Mitochondrial DNA Part. B 4 (1), 72–73. https://doi.org/10.1080/23802359.2018.1536464.

Tournadre, J., 2014. Anthropogenic pressure on the open ocean: the growth of ship traffic revealed by altimeter data analysis. Geophys. Res. Lett. 41, 7924–7932. https://doi.org/10.1002/2014GL061786.

Ushio, M., Fukuda, H., Inoue, T., Makoto, K., Kishida, O., Sato, K., et al., 2017. Environmental DNA enables detection of terrestrial mammals from forest pond water. Mol. Ecol. Resour. 17 (6), e63–e75. https://doi.org/10.1111/1755-0998.12690.

Whitehead, H., 2003. Sperm Whales: Social Evolution in the Ocean. University of Chicago Press, Chicago, IL, pp. 431.

Willis, P.M., Baird, R.W., 1998. Status of the Dwarf Sperm Whale. *Kogia* simus. with special reference to Canada. Can. Field Naturalist 112 (1), 114–125.

Printed and bound by CPI Group (UK) Ltd, Croydon, CR0 4YY
22/11/2024
01793009-0015